RNA Viruses and Neurological Disorders

This volume is an accessible introduction to RNA viruses and the infectious outcomes that they cause in the central nervous system (CNS). Chapters cover the major RNA viruses, their impact on the CNS, and the similarities and differences in pathological outcomes that can be observed. Neuroscientists, be they students, researchers, or clinicians, will benefit from the timely coverage provided.

Our understanding of viruses, and specifically RNA viruses and their pathological impact, is rapidly evolving. For example, the close molecular interaction of viruses with the CNS cell types in the human host is poorly understood. Readers can use the book to understand clearly the cellular and molecular mechanisms governing pathological outcomes of RNA virus infection in the cell types of the human brain (e.g., neurons, endothelial cells, astrocytes, resident immune cells) based on summarized case studies, and gain insight into how cell type-specific defects affect brain function and cause poor clinical outcomes.

The book is aimed primarily at neuroscience students and postgraduates wishing to learn about virology and professionals who are interested to learn more about virus-associated neuropathology. A basic knowledge of cell and molecular biology is assumed; however, readers across the disciplines of science, technology, engineering, and mathematics will find this topical and timely publication of value.

Sabyasachi Dash, MSc, PhD, is a Senior Scientist in the Exploratory Research Division at Alexion-AstraZeneca Rare Diseases located in New Haven, Connecticut. His work primarily focuses on drug discovery efforts for central nervous system disorders that are rare yet present a high unmet need in the society given their lack of understanding of the disease biology. He completed his postdoctoral training in neurovascular biology in the Department of Pathology and Laboratory Medicine at the Weill Cornell Medical College of Cornell University, New York, New York. His research focused on the development of novel vasoprotective therapeutics for vascular and cerebrovascular diseases by elucidating novel mechanisms regulating blood–brain barrier function in the mammalian brain. His PhD thesis work investigated the molecular mechanisms governing cocaine-induced neurotoxicity and neuroinflammation in the presence of HIV infection, which is of high pathological relevance in patients addicted to drugs of abuse.

RNA Viruses and Neurological Disorders

Edited by

Sabyasachi Dash

Department of Pathology and Laboratory Medicine
Weill Cornell Medicine
New York, New York, USA
Alexion-AstraZeneca Rare Diseases
New Haven, Connecticut, USA

CRC Press
Taylor & Francis Group
Boca Raton London

CRC Press is an imprint of the
Taylor & Francis Group, an **informa** business

First edition published 2024
by CRC Press
2385 NW Executive Center Drive, Suite 320, Boca Raton FL 33431

and by CRC Press
4 Park Square, Milton Park, Abingdon, Oxon, OX14 4RN

CRC Press is an imprint of Taylor & Francis Group, LLC

ISBN: 978-1-032-25680-1 (hbk)
ISBN: 978-1-032-25956-7 (pbk)
ISBN: 978-1-003-28582-3 (ebk)

DOI: 10.1201/9781003285823

Typeset in Sabon
by Apex CoVantage, LLC

Dedication

The book is dedicated to my family, my grandparents, and my career mentors whose teachings have sparked in me the enthusiasm and vigor to discover the unknown, innovate, and pass on the knowledge for a better future.

The book is an ode to the countless professionals across clinical and biomedical sciences whose efforts have advanced our understanding on the impact of RNA virus pathogenesis in the brain. Our knowledge would be obscure without the willingness of the families of patients who consented to undergo clinical investigation and share the clinical findings, thus providing us better insights into the virus-brain interplay.

Contents

Preface

This book is the first of its kind to introduce RNA viruses and the infectious outcomes that they cause in the central nervous system (CNS). The book is primarily aimed at undergraduate students at all levels and postgraduates wishing to learn about virology, as well as professionals who are interested to learn more about virus-associated neuropathology. The book approaches the subject in a fundamental manner by discussing the major RNA viruses and their CNS outcomes. By doing so, several important commonalities and uniqueness among the pathological outcomes can be observed, which may provide insights into disease management and therapy. Our aim in writing this book has been to cover the breadth of this fascinating and important subject on the nervous system impact of RNA virus infections, which has reemerged as a hot topic in recent years. It is thus suitable for students and professionals who may be studying virology or cellular and molecular aspects of neuroscience to obtain a better understanding from both the virus (pathogen) and human (host) perspectives. A basic knowledge of cell and molecular biology is assumed; however, readers across disciplines of science, technology, engineering, and mathematics should be able to approach this book. Our understanding of viruses, that too, specifically on RNA viruses and their pathological impact, is rapidly evolving. Thus, in consideration, we have aimed to maintain a broad coverage of the impact of RNA viruses on our health and society. Our quest for knowledge and innovation has been relentless, although much remains to be learned. For instance, the close molecular interaction of viruses with the CNS cell types in the human host is poorly understood. We have made attempts to highlight and discuss such instances on a case-by-case basis. The recent COVID-19 pandemic enhanced the public perception of viruses as significant threats to humans and animals, which has been a catalyst for the planning and development of this book. RNA viruses have been a major perpetrator for causing global health pandemics leaving a generational impact on a society. Thus, there has never been a more important time than now to understand the neurological and mental health impact caused by the highly communicable RNA viruses. Hence, it is of utmost importance to share the lessons learned from viral pandemics to elevate the understanding among the upcoming generation of healthcare experts, scientists, innovators, and caregivers.

Acknowledgments

This work would not have been possible without the support of the publishing and editorial team at CRC Press/Taylor & Francis. I am especially indebted to my contributing authors, who have been supportive of this project and who have worked diligently and tirelessly sharing the vision to provide the best quality scientific and clinical information to make this scholarly project a success.

I am grateful to all of those with whom I have had the pleasure to work during the journey of this project either directly or indirectly. I would especially like to acknowledge the efforts from my academic and professional mentors; because of them I am able to dare this scholarly stride—a small-scale attempt to impart evidence-based education and learning in the form of this book.

Nobody has been more important to me in the pursuit of this project than the members of my family. Most importantly, I wish to thank my loving and supportive wife, Prajna, and our wonderful baby boy whose arrival has been a blessing, a source of spiritual energy, and unending inspiration.

Contributors

Umme Abiha
IDRP, Indian Institute of Technology
    All India Institute of Medical
    Science
Rajasthan, India

Ridhika Bangotra
School of Biotechnology
University of Jammu
Jammu, India

Reshma Bhagat
Department of Psychiatry
Washington University
St. Louis, Missouri

Conor J. Cremin
Department of Microbiology and
    State Key Laboratory for Emerging
    Infectious Diseases
University of Hong Kong
Hong Kong, China

Sabyasachi Dash
Department of Pathology and
    Laboratory Medicine
Weill Cornell Medicine
New York, New York
and
Alexion-AstraZeneca Rare Diseases
New Haven, Connecticut

Neha Goel
Genetics and Tree Breeding
Forest Research Institute
Dehradun, India

Snehasmita Jena
School of Biotechnology
KIIT University
Odisha, India

Anne Khodarkovskaya
Department of Pathology and
    Laboratory Medicine
Weill Cornell Medicine
New York, New York

Apoorv Kirti
School of Biotechnology
KIIT University
Odisha, India

Parveen Kumar
Academy of Scientific & Innovative
    Research
Ghaziabad, India
and
Pharmacology Division
CSIR–Indian Institute of Integrative
    Medicine
Jammu and Kashmir, India

Sudakshya Sucharita Lenka
School of Biotechnology
KIIT University
Odisha, India

Shaikh Sheeran Naser
School of Biotechnology
KIIT University
Odisha, India

Suraj P. Parihar
Wellcome Centre for Infectious
    Diseases Research in Africa
    (CIDRI–Africa)
Institute of Infectious Diseases
    and Molecular Medicine
    (IDM)
University of Cape Town
Cape Town, South Africa

Nandan Patel
All India Institute of Medical
    Science
Rajasthan, India

Sukh Sandan
Department of Zoology
University of Delhi
Delhi, India

Gifty Sawhney
Academy of Scientific & Innovative
    Research
Ghaziabad, India
and
Pharmacology Division
CSIR–Indian Institute of Integrative
    Medicine
Jammu and Kashmir, India

Amal Senevirathne
College of Veterinary Medicine
Chungnam National University
Daejeon, South Korea

Mohit Sharma
Postgraduate School for Molecular
    Medicine
Medical University of Warsaw
Warsaw, Poland
and
Małopolska Centre of Biotechnology
Jagiellonian University
Krakow, Poland

Mrutyunjay Suar
School of Biotechnology
KIIT University
Odisha, India

Amrit Varsha
Department of Pharmacy
King's College London
London, UK

Suresh K. Verma
School of Biotechnology
KIIT University
Odisha, India

Part I

UNDERSTANDING THE RNA VIRUSES THAT INFECT THE HUMAN BRAIN

Chapter 1

Structure and Classification of RNA Viruses

Amal Senevirathne

1.1 INTRODUCTION

Viruses are strict parasites incapable of reproducing progeny on their own. They do not demonstrate the complete features required to classify as living organisms. Thus, it is reasonable to identify them as infectious particles. For survival, viruses are completely dependent upon the host's cellular machinery for replication purposes. Their genomes are made from deoxyribonucleic acid (DNA) or ribonucleic acid (RNA) (1,2), which are shielded in a protein coat known as a capsid. Because of their complete dependency on the host's cellular machinery, viruses are unable to perform essential life-sustaining tasks outside of their host cell. In other words, the protein synthesis process of viruses occurs on ribosomes burrowed into or highjacked from the host cell. They cannot synthesize or store the essential energy source, adenosine triphosphate (ATP), and acquire other essential components such as amino acids, nucleotides, lipids, and enzymes from the host cell via parasitizing the host cell. Owing to these reasons viruses are far too primitive to be considered as any sort of living thing. Therefore, recognizing viruses as a form of proto life is still debatable.

RNA viruses, like any other form of virus, can be classified based on size, shape, genome chemical makeup and structure, and mode of replication. Many filamentous and pleomorphic viruses have nucleocapsids with helical shapes (3). A helical array of capsid proteins known as protomers are wrapped around a helical filament of nucleic acid to form helical nucleocapsids. In contrast, numerous "spherical" viruses have nucleocapsids with an icosahedral form. Capsomeres are morphologic subunits of the icosahedron, and their number and arrangement are important for identification and classification (4). Furthermore, many viruses contain an outer membrane that performs a pivotal role in cellular adsorption and attachment (5).

DOI: 10.1201/9781003285823-1

1

1.2 MORPHOLOGY OF RNA VIRUSES

The morphology of viruses forms a symmetrical organization. The symmetry can be either helical or icosahedral (6). During the replication of helical viruses, identical protein subunits are self-assembled into a helical array covering the nucleic acid that is also organized in a helical arrangement. Helical nucleocapsids are rigid, highly elongated, and flexible filaments (7). These filaments can be in rigid, flexible, naked, or enveloped forms, and the number of subunits per helical turn, width, and height of virus filaments is considered during classification. One common example of a helical RNA virus is the tobacco mosaic virus (Figure 1.1A) that infects tobacco and other members of the family Solanaceae (8).

The icosahedral virion, which is a member of the icosahedral or quasi-spherical structural class of viruses, is composed of 20 identical triangular faces. Each face is made up of three identical capsid protein units, which together make up 60 subunits; five of these subunits are symmetrically positioned to contact each of the 12 vertices, putting all the proteins in equal interaction with one another. One or more proteins, each of which encodes a single viral gene, make up the symmetric protein capsid, housing its genome. In most icosahedral viruses, structural protein chains are organized as oligomeric clusters known as capsomeres, which can be visualized in detail using modern electron microscopy techniques. This symmetric arrangement allows viruses to use a limited number of genes to encode all the information required to build a huge capsid. The capsid containing

A   **B**

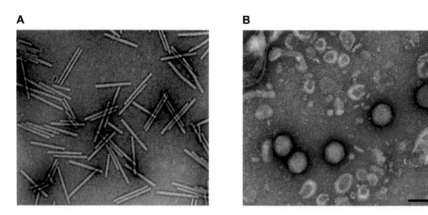

Figure 1.1 Morphological features of a filamentous RNA virus and an icosahedral RNA virus. (A) Electron micrograph of tobacco mosaic virus (TMV, 300 × 18 nm). Electron micrograph of TMV virions stained with uranyl acetate. (B) Electron micrograph of negatively stained poliovirus (magnification: 100,000 ×).

Sources: (A) Courtesy of Dr. J.N. Culver, University of Maryland Biotechnology Institute (1). (B) Courtesy of Dr. C. S. Goldsmith and S. E. Miller (2).

genetic material is also known as a nucleocapsid. In viruses with an envelope, the nucleocapsid is encased in a lipid bilayer and covered with a coating of glycoproteins. The nucleic acid is commonly connected to a protein known as a nucleoprotein, which performs an indispensable role during the replication of certain viruses (9). Poliovirus is a common example of an icosahedral virus (Figure 1.1B).

1.3 RNA VIRAL GENOMES

Seventy percent of all viruses are RNA viruses. Their genome structure demonstrates remarkable diversity. This diversity is partly influenced by the error-prone mechanism of RNA replication that incorporates mutations into the genome frequently. The mutation rate of RNA viruses is determined to be 1 in 10,000 nucleotide incorporations (10^{-4}), permitting tremendous diversity of progenitor viral particles that allow them to select and propagate highly host-adapted strains. The RNA genome of viruses can be single stranded (ss) or double stranded (ds) and be distributed as a single segment or multiple segments. A size comparison, genomic organization, and presence or absence of the outer membrane of a few selected human viruses are demonstrated in Figure 1.2.

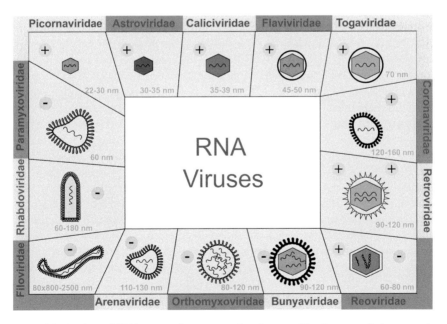

Figure 1.2 Families of RNA viruses known to infect humans. The features of viruses, such as shape, size, envelope presence or absence, RNA genome, + sense strand, and − antisense strand are indicated.

1.4 BASIC PROPERTIES OF RNA VIRUSES

RNA viruses are viruses that contain ribonucleic acid (RNA) as the genetic material. According to the International Committee on Taxonomy of Viruses (ICTV) (2), RNA viruses occupy *Group III, Group IV,* or *Group V* following the Baltimore classification system (10). The classification excludes RNA viruses which consist of deoxyribonucleic acid (DNA) intermediates that are known as retroviruses, which contain members of *Group VI* according to the Baltimore classification. All RNA viruses create a realm called *Riboviria* and all members contain RNA-dependent RNA polymerase as a major protein essential for the virus's life cycle. The realm *Riboviria* includes a majority of RNA viral families with a few exceptions such as *Asunviroidae, Pospiviroidae*, and the genus *Deltavirus*. The replication mechanism of all RNA viruses is dependent upon the virally encoded protein, RNA-dependent RNA polymerase (RdRp), which is one of the earliest proteins that starts its function (11). RdRp uses RNA viral genome as a template for the synthesis of a nascent RNA molecule. During the replication process, the RNA virus generates at least three types of RNA molecules that include RNA viral genome, a copy of the genome, and mRNA that undergo a translation process, which is occasionally processed by sub-genomic mRNA. Virally encoded RdRp is indispensable for these key biological functions that ensure the existence of RNA viruses.

For the synthesis of the viral genome, RdRp associates with several other proteins collectively known as replicase complex (12). The collective function of proteins associated with the replicase complex ensures the production of infectious viral genomes. A few of the key proteins associated with replicase complex include ATPase, which supplies energy; RNA-helicases, which ensure the unwinding of base-paired RNA structures; and proteases, while several other host-encoded proteins facilitate the replication process, and several other proteins that have unknown functions. In addition to viral proteins, there are host cell–associated proteins, such as polypyrimidine-tract-binding proteins (PTB), poly-A-binding proteins (PABP), heterogeneous nuclear ribonucleoprotein (hnRNP) A1, and mitochondrial aconitase (m-aconitase) in that it might interact with *cis*-acting elements during the replication process. The number of proteins associated with the formation of replication complex could differ among various RNA viral families. Evolutionary studies suggest proteins such as RdRp can be evolved from a common viral ancestor due to a high level of structural conservation. It is also notable that the biochemical machinery governing genome synthesis is not identical to the machinery of mRNA synthesis. Therefore, proteins required for viral mRNA synthesis are sometimes identified as *transcription complexes*. As a versatile protein molecule RdRp is involved in the transcription process of some viruses, where it consists of methylase activity that can be involved in the capping process of nascent mRNA while producing polyadenylated mRNAs, RdRp may "stutter" at poly U tracts.

1.5 CLASSIFICATION OF RNA VIRUSES

1.5.1 Positive-Stranded RNA Viruses

RNA viruses can be categorized into different categories according to the type of RNA that makes up their genomes. Positive-stranded RNA viral genomes consist of RNA molecules that could directly act as mRNA that could immediately translate into proteins (13). Upon release of the RNA genome, ribosomes are assembled on the genome and initiate viral protein synthesis. Initial proteins that are synthesized by the virus are the proteins that are essential for mRNA synthesis and synthesis of its genome (Figure 1.3). The RNA genome present in these viruses is single stranded

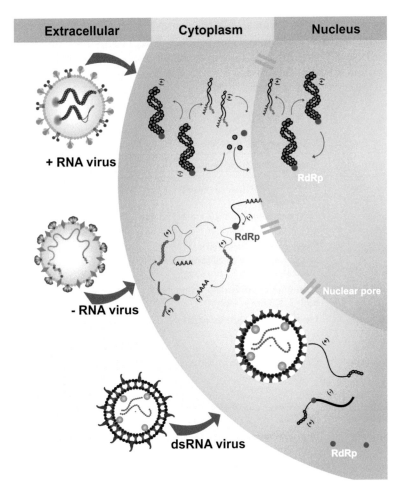

Figure 1.3 Major replication strategies present in RNA viruses.

and can undergo capping and polyadenylation. Thus, the viral genome of positive-stranded RNA viruses consists of two main functions, that is to provide a template for the synthesis of copies of the viral genome and serve as mRNA that carries out protein synthesis. In other words, the genome of positive-stranded RNA viruses is infectious. Whether it is chemically synthesized or purified, viral RNA remains infectious. Upon infection, positive-stranded RNA viruses modulate host cell membranes to facilitate viral replication by creating membrane complexes and replication scaffolds (14). The RdRp of positive-strand RNA viruses remains a non-structural protein. The expression of RdRp only occurs after a successful infection. The RdRp and other essential early proteins are encoded by the invading RNA genomes followed by a direct translation process. The resulting polyprotein is cleaved by virally encoded proteases. Some viruses produce a single long polyprotein (e.g., flaviviruses, picornaviruses). Viruses such as coronaviruses, arteriviruses, and togoviruses produce RdRp-containing polyproteins using whole-genome mRNA. However, structural proteins are encoded by subgenomic mRNA (Table 1.1).

Table 1.1 Features of Positive-Sense RNA Viruses—ICTV Classification

Virus Family	Phenotypic Features	Genome
Alphaflexiviridae	Flexuous filaments, 12–13 nm × 470–800 nm	Single, linear ss, 5.5–9.0 kb
Benyviridae	Non-enveloped rod, 20 nm × 390 nm	At least two segments, 4.6–7.0 kb
Bromoviridae	Spherical or quasi-spherical (26–35 nm diameter) or bacilliform (18–26 nm diameter, 30–85 nm long)	Three segments, linear, approx. 8 kb
Closteroviridae	Non-enveloped, flexible filaments of 650–2,200 nm × 12 nm	Mono- or bi-segment 13–19.3 kb
Endornaviridae	No true virions are associated	Monocistronic single-stranded RNA of 9.7–17.6 kb
Hadakaviridae	Capsidless (no known virions)	Multi-segmented, linear, 0.9–2.5 kb segments totaling 14–15 kb
Hypoviridae	Capsidless virus unable to form rigid particles	Unsegmented RNA, linear, 9.1–12.7 kb
Marnaviridae	Non-enveloped, 22–35 nm virions with four structural proteins	Non-segmented RNA, 8.6–9.6 kb
Nodaviridae	Non-enveloped spherical, 25–33 nm in diameter, with or without surface projections	Bipartite, ss, 3.1 kb (RNA1) and 1.4 kb (RNA2) with 5′-terminal caps but without poly (A) tails

Virus Family	Phenotypic Features	Genome
Polycipiviridae	Thought to be non-enveloped, 33 nm	Non-segmented, 10–12 kb
Roniviridae	Enveloped, bacilliform, 150–200 nm × 40–60 nm, helical nucleocapsid composed of the nucleocapsid protein (p20) surrounded by a lipid envelope containing two transmembrane glycoproteins (gp64 and gp116)	Non-segmented, single-stranded RNA of 26–29 kb containing five or six long open reading frames
Secoviridae	Non-enveloped, 25–30 nm	Mono- or bipartite RNA, 9–13.7 kb
Solinviviridae	Non-enveloped, 26–30 nm, apparent projections	Non-segmented, 10–11 kb
Virgaviridae	Non-enveloped, rod-shaped, 20 nm × 300 nm; except in members of the genus *Tobamovirus*, the particle length distribution is bi- or trimodal	Non-segmented in members of the genus *Tobamovirus* but multi-partite in other genera with segments separately encapsidated in two or three components
Arteriviridae	Pleomorphic, roughly spherical, 50–74 nm in diameter	Linear, 12.7–15.7 kb
Botourmiaviridae	Bacilliform (18 nm × 30–62 nm) with a 23.8 kDa coat protein; members of other genera are not encapsidated	Monopartite (2–5 kb) or tripartite (2.8 kb; 1.1 kb; 0.97 kb)
Caliciviridae	Non-enveloped with icosahedral symmetry, 27–40 nm in diameter	Linear, 7.4–8.3 kb, with a 5′-terminal VPg and 3′-terminal poly(A)
Dicistroviridae	Non-enveloped, 30 nm	Non-segmented, 8–10 kb
Flaviviridae	Enveloped, 40–60 nm virions with a single core protein (except genus *Pegivirus*) and two or three envelope glycoproteins	Non-segmented, 9.0–13 kb
Hepeviridae	Quasi-enveloped, 27–34 nm, single capsid protein	Monopartite, 6.4–7.2 kb
Iflaviridae	Non-enveloped, 22–30 nm	Non-segmented, 9–11 kb
Matonaviridae	Enveloped, 50–90 nm pleomorphic virions, spherical to tube-like, with a single capsid protein and two envelope glycoproteins	Non-segmented, 9.6–10 kb
Picornaviridae	Non-enveloped, 30–32 nm virions comprising 60 protomers	Non-segmented RNA with a poly(A) tail, 6.7–10.1 kb
Potyviridae	Non-enveloped, flexuous, and filamentous capsid, 650–950 nm × 11–20 nm, a single core capsid protein	Single-stranded, usually monopartite RNA (bipartite in genus *Bymovirus*), 8.2–11.5 kb
Sarthroviridae	Spherical, approx. 15 nm in diameter	Linear positive-sense ssRNA with a 3′-poly(A) tail

(Continued)

Table 1.1 Features of Positive-Sense RNA Viruses—ICTV Classification (Continued)

Virus Family	Phenotypic Features	Genome
Solemoviridae	Non-enveloped icosahedral, T = 3 symmetry, 20–34 nm in diameter, comprising 180 molecules of capsid protein	Non-segmented, with 5′-terminal VPg, no poly(A) tail, 4–6 kb
Togaviridae	Enveloped, 65–70 nm spherical, a single capsid protein and three envelope glycoproteins	Non-segmented, 10–12 kb
Yadokariviridae	Capsidless per se, trans-encapsidated into non-enveloped spherical virions, 33–50 nm in diameter	Non-segmented linear (+) RNA of 3.6–6.3 kb

1.5.2 Negative-Sensed RNA Viruses

Negative-sensed RNA viruses are classified into the family *Bunyaviridae*. This family includes both the negative-sensed RNA viruses and their close relative, ambisense RNA viruses. For negative-sensed RNA viruses, the first cellular event that occurs after cell penetration is transcription (15). The transcription process is undertaken by RdRp and other associated proteins that enter the cell along with the viral particle (Figure 1.3). Two to four proteins are associated with the transcription and replication complex, which can interact with the viral genome via interaction with RNA-binding nucleocapsid proteins (e.g., SARS-CoV-2 virus) (16). For this reason, purified or synthesized viral genome is not infectious and translatable. To begin with viral genome replication, the mRNA must be synthesized and translated. Due to negative-sensed RNA viruses containing RdRp within the virion, lysed virus particles under appropriate conditions could synthesize viral mRNA in vitro. But genomic RNA cannot be synthesized under in vitro conditions. By convention, the genome orientation of negative-sensed RNA viruses is depicted as 3′ to 5′ (left to right), opposite to the norm for depicting nucleic acids (Table 1.2).

1.5.3 Ambisense RNA Viruses

Molecular machinery associated with ambisense RNA viruses is closely related to negative-sensed RNA viruses. It can be considered a modified version of negative-sensed RNA viruses (17). Upon fusion of these viruses into a host cell, the virus's envelope is fused with the host cell membrane and releases its nucleocapsid into the host cell. The nucleocapsid remains intact and will not release viral RNA. Since the ambisense viral RNA is non-translatable, the first event will be transcription and the synthesis of positive-sensed RNA (antigenome). The required viral enzymes and proteins such as synthetases are associated with the nucleocapsid. Like negative-sense RNA viruses, the genome of ambisense RNA viruses is not infectious in purified RNA form. After synthesis of positive-sense RNA, it encodes mRNA that undergoes translation and

Table 1.2 Features of Negative-Sense RNA Viruses—ICTV Classification

Virus Family	Phenotypic Features	Genome
Arenaviridae	Enveloped, pleomorphic virions 40–200 nm, trimeric surface spikes	Two or three single-stranded, usually ambisense coding arrangement, RNA molecules called small (S), medium (M), and large (L)
Aspiviridae	Non-enveloped, naked filamentous nucleocapsids about 3 nm, form kinked circles of at least two different contour lengths (700 nm and approx. 2000 nm); pseudo-linear duplex structures are approx. 9–10 nm in diameter	Negative-sense, segmented RNA (three or four segments), 11.3–12.5 kb
Filoviridae	Enveloped, variously shaped, with a single nucleocapsid or polyploid	Non-segmented, linear, approx. 15–19 kb
Mymonaviridae	Enveloped, filamentous, 25–50 nm × 1,000 nm	Single-molecule, linear, 10 kb
Nyamiviridae	Enveloped, spherical particles, approx. 100–130 nm	Non-segmented or bisegmented, up to 13.3 kb
Peribunyaviridae	Enveloped spherical or pleomorphic virions, 80–120 nm	Three single-stranded, encode the nucleocapsid protein (N; S segment), envelope glycoproteins (Gn and Gc; M segment), and L protein that functions as the RNA-directed RNA polymerase and endonuclease (L; L segment)
Rhabdoviridae	Bullet-shaped or bacilliform particle 100–430 nm × 45–100 nm, helical nucleocapsid surrounded by a matrix layer and a lipid envelope; some have non-enveloped filamentous virions	Single-stranded RNA of 10.8–16.1 kb (unsegmented or bisegmented)
Artoviridae	Enveloped, spherical particles, 100–130 nm	Unsegmented, approximately 12 kb
Bornaviridae	Enveloped, spherical virions 90–130 nm	Linear, non-segmented, about 9 kb with three transcription units and at least six open reading frames (ORFs)
Fimoviridae	Approximately spherical and enveloped, diameter 80–100 nm	Four to eight segments of negative-sense ssRNA (12.3–18.5 kb in total)

(Continued)

Table 1.2 Features of Negative-Sense RNA Viruses—ICTV Classification (Continued)

Virus Family	Phenotypic Features	Genome
Nairoviridae	Enveloped, spherical virions 80–120 nm, heterodimer surface spikes	Three single-stranded, negative-sense RNA molecules, S, M, and L of approx. 2 kb, approx. 5 kb, and approx. 12 kb, respectively
Paramyxoviridae	Enveloped, pleomorphic (mostly spherical), 300–500 nm	Non-segmented, 14.6–20.1 kb
Pneumoviridae	Enveloped, both spherical and filamentous virions with a helical ribonucleoprotein (RNP) core	Unsegmented RNA genomes, 13.2–15.3 kb

produces proteins necessary for viral genome replication. Each mRNA acts as monocistronic mRNA, which will be translated only to a single protein, but not as a polyprotein. mRNA synthesis occurs within the nucleocapsid, and the nascent mRNA will be expelled to the cytoplasm for translation. The distinction of ambisense RNA viruses relies on the fact that these viruses use both genome and antigenome will serve as a template for mRNA synthesis.

1.5.4 Key Features of Negative-Sense and Ambisense RNA Viruses

The mRNA generated by both negative and ambisense mechanisms differ from their templates. These mRNAs do not contain promoter elements required for encapsidation or replication as genome or antigenome. These mRNA may not undergo encapsidation or do not synthesize their complementary strands. Second, such mRNA is capped and polyadenylated to facilitate their role as mRNAs, but the genomic or antigenomic RNA is not. The viruses belonging to the families *Bunyaviridae, Orthomyxiviridae*, and *Arenaviridae* consist of a 5′ overhang that is not present in genomic or antigenomic RNA. Most negative-sense RNA viruses and ambisense RNA viruses replicate in the cytoplasm. However, viruses such as influenza viruses and bornaviruses replicate in the nucleus, permitting them a unique opportunity to access host cellular-splicing enzymes.

1.5.5 Double-Stranded RNA Viruses

Double-stranded RNA (dsRNA) viruses belong to a large group of viruses forming 12 families. Most dsRNA viruses do not possess envelopes but form layers of concentric icosahedral capsids. Among these viral families, the family *Reoviridae* forms the largest family with the highest diversity. These viruses contain segmented genomes usually with 11–12 segments. They are non-enveloped but contain two to three layers of the icosahedral capsid. The genome segments are associated with the innermost capsid layer. Transcription of dsRNA viruses occurs within the capsid shell, and the resulting mRNA is exported from the capsid via capsid pores located at capsid vertices (Figure 1.3).

1.6 STRUCTURE OF RNA VIRAL GENOMES

All RNA viruses share common genomic features; for example, their genomes contain single or multiple open reading frames. Each open reading frame encodes a protein that performs a specific function for viral infection and replication processes. RNA viral genomes consist of genomic regions that do not code for any protein. These non-coding regions are often highly conserved within each viral family and thus can anticipate the significant necessity for the life cycle of an RNA virus. Such non-coding regions occasionally are referred to as untranslated regions (UTRs). The nucleotide sequence of such regions might play a role in determining the sequence-specific folding pattern whose function may be critical for the life cycle of a virus. Non-coding regions of RNA viral genomes occur at both 5′ and 3′ termini, whereas the length of non-coding regions (NCRs) can vary from a few hundred nucleotides (e.g., picornavirus) to short sequences (e.g., flavivirus). On some occasions, nucleotide sequences at the 5′ and 3′ ends may be complementary to each other, allowing the viral genome to circularize. In addition, these NCRs play a role in genome replication and are often sufficient for direct genome replication. In the viral genome, special sequences such as promoter regions are there whose function is to provide recognition sites for virally encoded proteins such as RdRp. RdRp initiates transcription upon interaction with the RNA genome via specific recognition sequences located at the genome termini. For experiment purposes, scientists generate RNA genomes that may be absent with certain elements such as virulence-related genes. However, to initiate transcription, the essential protein RdRp must be supplied externally, encoded by the modified genome, or stably expressed by the host cell. The genome sequences that are essential for the direct replication of RNA often consist of simple confirmation, and thus can be connected with other RNA sequences to undertake RNA replication.

Every RNA genome has promoters that control the synthesis of subgenomic mRNAs. These promoter sequences, although they may be brief, offer a way to position the RdRp to internal genomic locations. Additionally, there can be specific RNA sequences that indicate polyadenylation. The specific sequences essential for all the mechanisms related to RNA genome replication and transcription are embedded in the RNA genome itself. Genomes of some RNA viruses are highly structured; for example, internal ribosome entry sites (IRES) of members of *Picornaviridae*. In these genomes, ribosome assembly occurs on platforms provided by IRES sequences, which are highly based paired regions. The unwinding of these structured regions requires virally encoded specialized proteins such as RNA helicases. Without the assistance of RNA helicases, these highly base-paired regions can halt genomic replication. Thus structural unwinding and the role of helicases are indispensable processes.

1.7 MAJOR EVENTS INVOLVED IN VIRAL RNA SYNTHESIS

The enzyme RdRp performs a pivotal role in all steps of the viral RNA genome replication and transcription processes. RdRp has independently evolved for each viral family to recognize the 3′ end of viral genomes, ensure functionality and integrity of copied RNA, and also evade recognition of RNA genome from the host innate immune system. RNA viruses utilize a variety of strategies for priming and regulating RNA synthesis and also the capping and polyadenylation processes of transcribed mRNA.

Deployment of DNA-dependent RNA polymerases in eukaryotic cells is primer independent. Specialized sequences known as promoters independently direct RNA polymerases to the site of initiation located on DNA genomes. The particular promoter sequences will not be transcribed. Compared to viral genomes, promoter regions of eukaryotic cells are long and complex. In contrast, RNA viruses use various mechanisms to utilize RdRp during the replication and transcription of their RNA genomes. During genome synthesis, it is important to maintain the integrity of the viral genome without significant mutation or the loss of genomic information. In RNA viruses, synthesis of RNA stretches may occur with and without primers. During a primer-independent genome, replication occurs with the involvement of RdRp at the template by employing a sequence of nucleotide triphosphates (NTPs), thus the very first NTP compensate for a role of a primer. On other occasions, short oligonucleotide primer sequences are first synthesized and used as primers that function in a prime-realigned fashion. In Bunyaviruses and arenaviruses, the primer sequence extends to the 5′ end of the copy genome, thus non-templated nucleotides can be identified. In addition, short hairpin structures also looped back to form a similar effect of a primer allowing genome synthesis. In this method the template itself provides a priming effect that is essential for genome replication. The family *Picornaviridae* uses a protein-mediated priming process where the RdRp, RNA template, and GTP-linked tyrosine hydroxyl group act as the initial nucleotide by facilitating RNA genome extension. In the case of influenza viruses (*Orthomyxoviridae*), mRNA synthesis will occur by burrowing host cellular mRNA for priming purposes. However, their genome synthesis requires de novo synthesized RNA primers.

1.8 MRNA CAPPING AND POLYADENYLATION

The majority of viruses convert their mRNAs into proteins using the mRNA cap–dependent cellular translation machinery. For many virus families, adding a cap structure to the 5′ end of mRNA is consequently a necessary step in the replication process. The cap also shields viral RNA from cellular nuclease breakdown and hinders the detection of viral RNA by innate immune mechanisms. Viral RNAs either use virus-encoded capping enzymes, cellular capping enzymes, or a mechanism known as "cap snatching" to steal the cap from cellular mRNA to form their cap structure. In influenza viruses, the priming process of viral mRNA uses fragments of cellular mRNA and also provides a methyl-G cap. Furthermore, the RdRp of many RNA viruses possesses methylase activity, hence it can be involved in the capping process of 5′ termini of nascent RNA. Recently, the structural and functional characteristics of several viral enzymes participating in this process have been identified. Also, there are RNA virus families that do not use capped mRNA; for example, *Picornaviridae*. RNA viruses use several different methods to polyadenylate their mRNA; for instance, poly-A tracts that are encoded in the genome are used by picornaviruses. The order *Mononegavirales* of negative-strand RNA viruses employ a stuttering mechanism to create long poly-A tracts from short poly-U tracts. Other RNA viruses' mRNAs are not polyadenylated. The mRNA of the flaviviruses has a brief hairpin at the 3′ termini.

1.9 REGULATION OF GENOME SYNTHESIS AND TRANSCRIPTION

All RNA viruses strictly regulate the amount of genome required, the amount of copy genome, and the quantity of mRNA to avoid wastage of resources. If an RNA virus had to create a copy genome for every genome, the synthesis would be a wasteful process. Instead, a copy genome can involve in the synthesis of many genomes. RNA viruses use different structures at 5′ and 3′ termini at the genome and copy the genome for this objective. Also, some viruses use different replicase complexes to synthesize RNA genomes and copy genomes. Here, the replicates may be regulated by proteolytic activity. The regulation of mRNA may occur by varying affinities for transcription initiation complexes towards variable promoter sequences. The negative-stranded RNA viruses present in the order *Mononegavirales* possess a sequential expression of mRNA from the 3′ to 5′ direction of their infecting genome, hence, there is a gradient of gene expression with more at the 3′ terminus and less at the 5′ terminus. The particular style of mRNA expression can be seen in RNA viruses belonging to families *Filoviridae, Paramyxoviridae, Rhabdoviridae, Pneumoviridae,* and *Bornaviridae*.

In addition to the major viral classification, most viruses exist as quasispecies. Groups of quasispecies are closely related but genomes are not identical. This is not only specific to RNA viruses but common to other types of viruses such as retroviruses. Human immunodeficiency virus and poliovirus are some examples of quasispecies. The existence of viruses as quasispecies may be advantageous for their survival as a population, rather than individual viruses. Studies have identified that any single genome of a population of quasispecies does not replicate better than the group as a whole. More often, if a genome is separated from the population and allows them to cause infection and replication, its replicability is often poor than the whole population. The exact reason for quasispecies occurrence is not clear. However, there can be advantages of being quasispecies to survive in a hostile environment such as animal cellular environments.

One major mechanism of quasispecies occurrence is dependent upon the fidelity of RdRp. The fidelity of RdRp is low for viruses that form quasispecies such as poliovirus. During the replication process, RdRp permits the accumulation of mutations, resulting in the abortion of transcription or continued transcription irrespective of the mutations. In this method, point mutations are integrated into the genome more frequently than transverse mutations, where a purine base is replaced with a pyrimidine base. The average rate of incorporation of point mutations is about 1/1000–1/100,000, whereas the incorporation of transverse mutation is about 1/1,000,000–1/10,000,000. Thus, after several replication cycles, plenty of mutations are incorporated into a viral genome of a population. The generation of quasispecies ensures that the population is better off than the individual entities of viral genomes.

1.10 CONCLUSION

The structure and classification of RNA viruses are based on the structural features of viral particles and the organization of their genomes. This enables us to identify the phylogenetic relationship of a virus, in relation to all the other viruses in the classification system. Classification encompasses clinical, biological, and evolutionary features to be compiled into a framework that connects all known RNA viruses under the broader classification of all known plant, animal, and prokaryotic viruses. Such understanding has practical importance when a novel virus is recognized or when a virus switches its traditional host range, to effectively trace its ancestral positions according to classification. Such information is not only important to find virus-controlling measures but also to propose frameworks for potential future outbreaks. Future progress will enhance the integration of the virus and cellular taxonomic systems and also have profound effects on the study of microbes, plants, animals, medicine, computation, and the environment. Finally, the social, economic, and individual effects of the epidemics and pandemics that have been and will continue to be brought on by viruses cannot be underestimated.

Abbreviations

ATP	Adenosine triphosphate
DNA	Deoxyribonucleic acid
dsRNA	Double-stranded RNA
GTP	Guanidine triphosphate

hnRNP	Heterogeneous nuclear ribonucleoprotein
ICTV	International Committee on Taxonomy of Viruses
IRES	Internal ribosome entry sites
m-aconitase	Mitochondrial aconitase
mRNA	Messenger ribonucleic acid
NCRs	Non-coding regions
NTP	Nucleotide triphosphate
PABP	Poly-A-binding proteins
PTB	Polypyrimidine tract-binding proteins
RdRp	RNA-dependent RNA polymerase
RNA	Ribonucleic acid
SARS-CoV-2	Severe acute respiratory syndrome coronavirus 2
UTR	Untranslated region

REFERENCES

1. Naz SS, Aslam A, Malik T. An overview of immune evasion strategies of DNA and RNA viruses. *Infection Disorder Drug Targets* [Internet] 2021 [cited 2023 Mar 7]; 21. Available from: https://pubmed.ncbi.nlm.nih.gov/33739247/

2. Walker PJ, Siddell SG, Lefkowitz EJ, Mushegian AR, Adriaenssens EM, Dempsey DM, Dutilh BE, Harrach B, Harrison RL, Hendrickson RC, et al. Changes to virus taxonomy and the statutes ratified by the International Committee on Taxonomy of Viruses. *Archives of Virology*, 2020; 165: 2737–2748.

3. Makarov VV, Kalinina NO. Structure and noncanonical activities of coat proteins of helical plant viruses. *Biochemistry* (Moscow) 2016; 81.

4. San Martín C. Transmission electron microscopy and the molecular structure of icosahedral viruses. *Archives of Biochemistry and Biophysics* 2015; 581.

5. Li W, van Kuppeveld FJM, He Q, Rottier PJM, Bosch BJ. Cellular entry of the porcine epidemic diarrhea virus. *Virus Research* 2016; 226.

6. Ng WM, Stelfox AJ, Bowden TA. Unraveling virus relationships by structure-based phylogenetic classification. *Virus Evolution* 2020; 6.

7. Louten J. Virus structure and classification. *Essential Human Virology* 2016; 19–29. doi: 10.1016/B978-0-12-800947-5.00002-8.

8. Liu C, Nelson RS. The cell biology of tobacco mosaic virus replication and movement. *Frontiers in Plant Science* 2013; 4.

9. Turrell L, Lyall JW, Tiley LS, Fodor E, Vreede FT. The role and assembly mechanism of nucleoprotein in influenza A virus ribonucleoprotein complexes. *Nature Communications* 2013; 4.

10. Koonin EV., Krupovic M, Agol VI. The Baltimore classification of viruses 50 years later: How does it stand in the light of virus evolution? *Microbiology and Molecular Biology Reviews* 2021; 85.

11. Tian L, Qiang T, Liang C, Ren X, Jia M, Zhang J, Li J, Wan M, YuWen X, Li H, et al. RNA-dependent RNA polymerase (RdRp) inhibitors: The current landscape and repurposing for the COVID-19 pandemic. *European Journal of Medicinal Chemistry* 2021; 213.

12. Tao YJ, Ye Q. RNA virus replication complexes. *PLoS Pathogens* 2010; 6.

13. Shi J, Luo H. Interplay between the cellular autophagy machinery and positive-stranded RNA viruses. *Acta Biochimica et Biophysica Sinica* (Shanghai) 2012; 44.

14. Hyodo K, Suzuki N, Okuno T. Hijacking a host scaffold protein, RACK1, for replication of a plant RNA virus. *New Phytologist* 2019; 221.

15. Ortín J, Martín-Benito J. The RNA synthesis machinery of negative-stranded RNA viruses. *Virology* 2015; 479–480.

16. Kang S, Yang M, Hong Z, Zhang L, Huang Z, Chen X, He S, Zhou Z, Zhou Z, Chen Q, et al. Crystal structure of SARS-CoV-2 nucleocapsid protein RNA binding domain reveals potential unique drug targeting sites. *Acta Pharmaceutica Sinica B* 2020; 10.

17. Nguyen M, Haenni AL. Expression strategies of ambisense viruses. *Virus Research* 2003; 93.

Chapter 2

RNA Viruses with Central Nervous System Tropism

Amrit Varsha

2.1 INTRODUCTION

It was only approximately 100 years ago that viruses were shown to be filterable and therefore distinct from bacteria that cause infectious disease. It was only about 60 years ago that the composition of viruses was described, and even more recently before they could be visualized as particles in the electron microscope. Within the last 20 years, however, the revolution of modern biotechnology has led to an explosive increase in our knowledge of viruses and their interactions with their hosts. Virology, the study of viruses, includes many aspects: the molecular biology of virus replication, the structure of viruses, the interactions of viruses and hosts and the diseases they cause in those hosts, the evolution and history of viruses and viral diseases, virus epidemiology, the ecological niche occupied by viruses and how they spread from victim to victim, and the prevention of viral disease by vaccination, drugs, or other methods. The field is vast, and any treatment of viruses must perforce be selective.

Human viruses that cause disease, especially epidemic disease, are not uniformly distributed across virus families, thus this treatment is not intended to be comprehensive. Nevertheless, we feel that it is important that the human viruses be presented in the perspective of viruses so that some overall understanding of this fascinating group of agents can emerge. Thus, we consider many nonhuman viruses that are important for our understanding of the evolution and biology of viruses. The biological diversity and rapid adaptive rate of RNA viruses have posed a great challenge for modern medical technologies. Also, the anthropogenic change of natural ecosystems and the continuous population growth have increased the rate of interspecies contact and transmission. As a result, such cross-species transmission of pathogens/viruses has resulted in global pandemics. The combination of molecular epidemiological and ecological knowledge of RNA viruses is therefore essential towards the proper control of current and emergent pathogens (1).

The use of vaccines has led to effective control of the most dangerous of the viruses. Smallpox virus has been eradicated worldwide by means of an ambitious and concerted effort, sponsored by the World Health Organization, to vaccinate all people at risk for the disease. Previously, both poliovirus and measles viruses were reported to have been eliminated from the Americas by intensive vaccination programs; however, recent Centers for Disease Control (CDC) estimates report the re-appearance of more than 1,000 cases of measles

DOI: 10.1201/9781003285823-2

infection across 31 states in the United States. There is hope that these two viruses will also be eradicated worldwide soon. Vaccines exist for the control of many other viral diseases including, among others, mumps, rabies, rubella, yellow fever, Japanese encephalitis, rotaviral gastroenteritis, and, very recently, papillomaviral disease, which is the primary cause of cervical cancer.

The dramatic decline in the death rate from infectious disease has led to a certain amount of complacency. Developed countries as well as developing countries suffer from viruses for which no vaccines exist at the current time. Human immunodeficiency virus (HIV), illustrated in Figure 4.7 (in Chapter 4), is a case in point. The persistence of viruses is in part due to their ability to change rapidly and adapt to new situations. HIV is the most striking example of the appearance of a virus that has recently entered the human population and caused a plague of worldwide importance. The arrival of this virus in the United States caused a noticeable rise in the total number of deaths.

In addition to the interest in viruses that arises from their medical and scientific importance, viruses form a fascinating evolutionary system. There is debate as to how ancient are viruses. Some argue that RNA viruses contain remnants of the RNA world that existed before the invention of DNA. All would accept the idea that viruses have been present for hundreds of millions of years and have helped to shape the evolution of their hosts. Viruses are capable of very rapid change, both from drift due to nucleotide substitutions that may occur at a rate 1 million-fold greater than that of the plants and animals that they infect, and from recombination that leads to the development of entirely new families of viruses. This makes it difficult to trace the evolution of viruses back more than a few millennia or perhaps a few million years. The development of increasingly refined methods of sequence analysis, and the determination of more structures of virally encoded proteins, which change far more slowly than do the amino acid sequences that form the structure, have helped identify relationships among viruses that were not at first obvious. The coevolution of viruses and their hosts remains a study that is intrinsically interesting and has much to tell us about human biology.

In virology, the *realm* is the highest taxonomic rank established for viruses by the International Committee on Taxonomy of Viruses (ICTV), which oversees virus taxonomy. Six virus realms are recognized and united by specific highly conserved traits.

The two realms associated with RNA viruses are:

- *Riboviria*, which contains all RNA viruses that encode RNA-dependent RNA polymerase and all viruses that encode reverse transcriptase.
- *Ribozyviria*, which contains hepatitis delta–like viruses with circular, negative-sense ssRNA genomes.

The rank of realm corresponds to the rank of domain used for cellular life but differs in that viruses in a realm do not necessarily share a common ancestor based on common descent, nor do the realms share a common ancestor. Instead, realms group viruses together based on specific traits that are highly conserved over time, which may have been obtained on a single occasion or multiple occasions. As such, each realm represents at least one instance of viruses coming into existence. While historically it was difficult to determine deep evolutionary relations between viruses, in the 21st century methods such as metagenomics and cryogenic electron microscopy have enabled such research to occur, which led to the establishment of *Riboviria* in 2018, three realms in 2019, and two in 2020. The virus family concept is fundamentally important in understanding the biological classification of viruses. By specifying the family to which a virus belongs, much can be inferred about its physical, chemical, and biologic properties and its evolutionary relationships including modes of gene expression. Virus families are designated with the suffix *-viridae*. Families are distinguished largely based on physiochemical properties, genome structure, size, morphology, and molecular processes. Viruses are microscopic obligate intracellular parasites which contain either an RNA or DNA genome surrounded by a protective protein coat called a capsid (2). Viruses can be classified based on their morphology, chemical composition, structure of the genome, and various modes of replication, as shown in Figure 2.1. The viral genome may either be single stranded (ss) or double stranded (ds) and is packaged in either a linear or circular arrangement. The entire genome may either constitute a single nucleic acid molecule (monopartite) or multiple copies of the nucleic acid molecule (multipartite). Such different types of genomes necessitate different replication strategies based on their genomic structure. RNA viruses are of two types: single-stranded RNA (ssRNA) and double-stranded RNA (dsRNA).

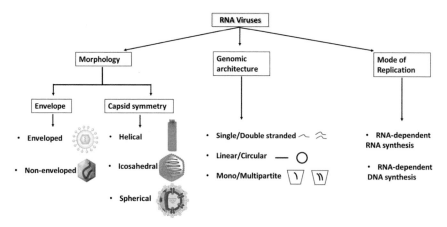

Figure 2.1 General classification of RNA viruses. RNA viruses can be categorized based on their genome architecture, shape, and mode of replication.

The single-stranded RNA viruses are further classified as (1) [+ssRNA], plus strand, or positive strand; and (2) [–ssRNA], minus strand, or negative strand. Viruses are also classified based on their morphology into two main categories: (1) enveloped, which have a lipid membrane (envelope); and (2) non-enveloped, which lack a membrane. Given the shape, helical morphology is seen in nucleocapsids of several filamentous and pleomorphic viruses. Helical nucleocapsids consist of a helical array of capsid proteins called protomers wrapped around a helical filament of nucleic acid. On the other hand, icosahedral morphology is characteristic of the nucleocapsids of several spherical viruses. The number and arrangement of the capsid proteins are useful in identification and classification of viruses.

2.2 CALICIVIRIDAE

Caliciviridae are small, round-structured viruses that bear resemblance to enlarged picornavirus. They are non-enveloped, positive-sense, monopartite, and single-stranded RNA that is packaged into an icosahedral, environmentally stable protein capsid (Figure 2.2). The family contains five genera (*Norovirus, Nebovirus, Sapovirus, Lagovirus,* and *Vesivirus*) that infect vertebrates including amphibians, reptiles, birds, and mammals. The RNA-dependent RNA polymerase (RdRp) replicates the genome of RNA viruses and can speed up evolution due to its error-prone nature. Studying calicivirus RdRps in the context of genuine virus replication is often hampered by a lack of suitable model systems. Enteric caliciviruses and RHDV (rabbit hemorrhagic disease virus) are notoriously difficult to propagate in cell culture; therefore, molecular studies of replication mechanisms are challenging. Nevertheless, research on recombinant proteins has revealed several unexpected characteristics of calicivirus RdRps. For example, the RdRps of RHDV and related Lago viruses possess the ability to expose a hydrophobic motif, to rearrange Golgi membranes, and to copy RNA at unusually high temperatures (3).

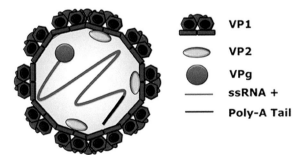

Figure 2.2 Structure of human norovirus.

Source: Figure reprinted with permission from (4).

Diseases associated with this family include feline calicivirus (respiratory disease), rabbit hemorrhagic disease virus (often fatal hemorrhages), and the Norwalk group of viruses (gastroenteritis). Caliciviruses are responsible for gastroenteritis in cruise ships and in school settings (Norovirus infection). The Norwalk virus, now called Norovirus, is the most common cause of epidemic nonbacterial gastroenteritis in the world. Gastroenteritis is an infection of the stomach or intestines. The virus was initially named due to an outbreak in an elementary school in 1968 in Norwalk, Ohio, but was renamed in 2002. Other, less known Noroviruses are rabbit hemorrhagic disease virus, Sapporo virus, and vesicular exanthema of swine virus. Common symptoms include diarrhea, vomiting, nausea, and stomach pain.

2.3 PICORNAVIRIDAE

These viruses are a group of non-enveloped +ssRNA viruses that infect vertebrates including fish, mammals, and birds. They are viruses that represent a large family of small, positive-sense, single-stranded RNA viruses with a 30-nm icosahedral capsid (Figure 2.3). The viruses in this family can cause a range of diseases including the common cold, poliomyelitis, meningitis, hepatitis, and paralysis. Picornaviruses constitute the family *Picornaviridae*, order *Picornavirales*, and realm *Riboviria*. There are 158 species in this family assigned to 68 genera. Notable examples are genera *Enterovirus* (including *Rhinovirus*, which causes common cold, and *Poliovirus*, which causes poliomyelitis), *Aphthovirus*, *Cardiovirus*, and *Hepatovirus*. Enteroviruses are named by their transmission route through the intestine (e.g., *enteric* means "intestinal"). Among this type of virus, Coxsackie B viruses are widely distributed and can cause meningitis (inflammation of the membranes that line

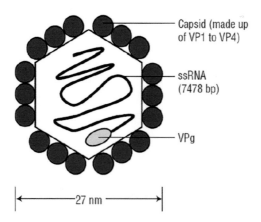

Figure 2.3 Example of *Picornaviridae* family of viruses (structure of poliovirus).

Source: Image reprinted with permission from (5).

the brain and spinal cord). The Coxsackie B viruses also cause paralysis due to the degeneration of central nervous system (CNS) neurons. Echoviruses cause nonspecific viral infections that can include meningitis, encephalitis, and paralysis. Picornavirus infections have been a challenging problem in human health. Genome organization of picornavirus is unique in having a long, heavily structured, multifunctional 5′ untranslated region, preceding a single open reading frame from which all viral proteins are produced. Within the 5′ leader, a region termed the internal ribosome entry site (IRES) regulates viral protein synthesis in a 5′-independent manner. The IRES element itself is a distinctive feature of the picornavirus mRNAs, allowing efficient viral protein synthesis in infected cells despite a severe modification of translation initiation factors induced by viral proteases that lead to a fast inhibition of cellular protein synthesis. Picornavirus IRES elements are strongly structured, bearing several motifs, phylogenetically conserved, which are essential for IRES activity. Together with the RNA structure, RNA-binding proteins play an essential role in the activity of the IRES element, having a profound effect on viral pathogenesis (6). Picornavirus IRES are amongst the most potent elements described so far. However, given their large diversity and complexity, the mechanistic basis of its mode of action is not yet fully understood (7).

2.4 *TOGAVIRIDAE* **AND** *FLAVIVIRIDAE*

These enveloped virions are spherical, 60 to 70 nm in diameter with a positive-sense, monopartite, single-stranded RNA genome about 11.7 kilobases long (8). The lipid-containing envelope has two (rarely three) surface glycoproteins that mediate attachment, fusion, and penetration. The icosahedral nucleocapsid contains capsid protein and RNA (Figure 2.4). Virions

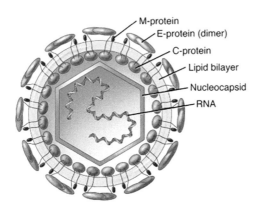

M-protein
E-protein (dimer)
C-protein
Lipid bilayer
Nucleocapsid
RNA

Figure 2.4 Representative architecture of the West Nile virus, a member of the Togavirus family.

Source: Image reprinted with permission from (9).

mature by budding through the plasma membrane. Many of the important zoonotic arboviruses belong to the families *Togaviridae* and *Flaviviridae*. Viruses that maintain transmission cycles between vertebrate animal reservoirs as main amplifying hosts and insects as primary vectors are known as Arboviruses (arthropod-borne viruses). Arboviruses must replicate in the arthropod vectors, such as mosquitoes, ticks, midges, or sandflies, prior to transmission. Arboviruses is an informal name for any virus that transmitted by arthropod vectors (mosquitos, fleas, and ticks). Sometimes these viruses are distinguished between those transmitted by mosquitoes or ticks. The La Crosse virus (also called California virus) causes encephalitis, mostly in children. Eastern equine encephalitis virus predominates in the eastern United States and affects mainly young children and people older than 55. West Nile virus causes the West Nile fever or encephalitis, which was originally present only in Europe and Africa but spread throughout the United States. This encephalitis affects mainly older people. The St. Louis virus also causes encephalitis, mostly in urban areas of central and southeastern states of the United States but also in western states and mostly affects older people. Western equine encephalitis is caused by the virus of the same name throughout the United States, but cases have not been reported since 1988.

2.4.1 *Flaviviridae*

These family of viruses were once considered to be of the *Togaviridae* family; however, they are now designated as members of a separate family, *Flaviviridae*. The *Flaviviridae* are a family of positive, single-stranded, enveloped RNA viruses (Figure 2.5). They are found in arthropods (primarily ticks and mosquitoes) and can occasionally infect humans. Among virus transmitted by ticks that cause encephalitis is the Powassan virus, which affects the Great Lakes region of the United States and Canada. The Powassan virus is spread by the deer tick, which also transmits Lyme disease. Tick-borne encephalitis occurs in northern Asia, Russia, and Europe. The infection usually causes a mild flulike illness that clears up within a few days. Colorado tick fever, as

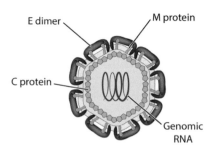

Figure 2.5 Structure of dengue virus.

Source: Image reprinted with permission from (10).

its name indicates, occurs in high altitude areas of the western United States and Canada. This virus can cause meningitis or encephalitis. Many of the important zoonotic arboviruses belonging to the family of *Flaviviridae* are dengue virus, yellow fever virus, Japanese encephalitis, West Nile viruses, and Zika virus. Other *Flaviviruses* are transmitted by ticks and are responsible for encephalitis and hemorrhagic diseases are tick-borne encephalitis (TBE), Kyasanur forest disease (KFD), Alkhurma disease, and Omsk hemorrhagic fever.

2.5 CORONAVIRIDAE

This family of viruses are enveloped positive single-strand RNA (+ssRNA) viruses that infect amphibians, birds, and mammals. The group includes the subfamilies *Letovirinae* and *Orthocoronavirinae*. The members of the latter are known as Coronaviruses, named for their "sunlike" shape observed in the electron microscope. It uses RNA molecules to encode their genes like influenza viruses, HIV, rhinoviruses (the common cold), and SARS-CoV-2, the virus that causes COVID-19. The latter infects mammals and birds and is closely related to the viruses causing the earlier SARS (severe acute respiratory syndrome) and MERS (Middle East respiratory syndrome) outbreaks. The coronavirus particles are organized with long RNA polymers tightly packed into the center of the particle, and surrounded by a protective capsid, which is a lattice of repeated protein molecules referred to as coat or capsid proteins (Figure 2.6). In coronavirus, these proteins are called nucleocapsids

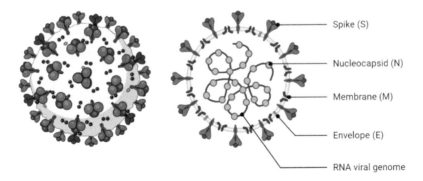

Figure 2.6 Structure of coronavirus. The coronavirus constitutes long RNA polymers that are tightly packed into the core of the viral particle and surrounded by a protective capsid. This capsid serves as a lattice of repeated protein molecules referred to as viral coat or capsid proteins also known as the nucleocapsid (N). The outer membrane envelope is made of lipids (fats) with viral proteins embedded. The virus membranes are derived from the cells in which the virus was assembled and budded out. Each viral particles contains specific viral proteins, including the spike (S), membrane (M), and envelope (E) proteins.

Source: Figure reprinted with permission from (11).

(N). The coronavirus core particle is further surrounded by an outer membrane envelope made of lipids with proteins inserted. These membranes derive from the cells in which the virus was last assembled but are modified to contain specific viral proteins, including the spike (S), membrane (M), and envelope (E) proteins. A key set of the proteins in the outer membrane project out from the particle and are known as spike proteins. It is these proteins that are recognized by receptor proteins on the host cells, which will be infected. Coronaviruses primarily infect human lung cells through a receptor for an enzyme called angiotensin-converting enzyme 2 (ACE2). ACE2 is a member of the family of angiotensin-converting enzymes that includes ACE. As the first step leading to viral infection, the virus spike protein recognizes and binds to the ACE2 receptor. The virus infects lung cells by binding ACE2 on the cell surface, which leads to leukocyte infiltration, increased permeability of blood vessels and alveolar walls, and decreased surfactant in the lung, causing respiratory symptoms. The aggravation of local inflammation causes cytokine storm, resulting in systemic inflammatory response syndrome (12,13).

2.6 RETROVIRIDAE

The family of retroviruses constitutes a large family of viruses that predominantly infect both human and animal vertebrates. They are positive-stranded enveloped RNA viruses that reverse transcribe their RNA into a DNA intermediate during viral replication, hence the name "retroviruses" (Figure 2.7). Retroviral infections can cause a wide spectrum of diseases ranging from malignancies to immune deficiencies and neurologic disorders (14–18). However, most retroviral infections occur without any detectable deleterious effect to the host. Examples of retroviruses are the human immunodeficiency virus (HIV) and the human T cell leukemia viruses (HTLVs). First brought to scientific attention as infectious cancer-causing agents nearly 80 years ago, retroviruses are popular in contemporary biology for the following reasons. (a) The virus life cycle includes several events in particular: reverse transcription of the viral RNA genome into DNA, orderly integration of viral DNA into host chromosomes, and utilization of host mechanisms for gene expression in response to viral signals, which are broadly informative about eukaryotic cells and viruses. (b) Retroviral oncogenesis usually depends on transduction or insertional activation of cellular genes, and isolation of those genes has provided the scientific community with many of the molecular components now implicated in the control of normal growth and in human cancer. (c) Retroviruses include many important veterinary pathogens and two recently discovered human pathogens, the causative agents of the acquired immunodeficiency syndrome (AIDS) and adult T cell leukemia/lymphoma. (d) Retroviruses are genetic vectors in nature and can be modified to serve as

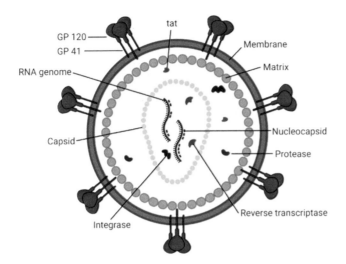

Figure 2.7 Structure of HIV virus. The viral envelope proteins, GP41 and GP120, surround the host-derived membrane surface from where the virions assembled, budded, and exited, which is lined internally with a layer of matrix protein. Inside the virion are viral proteins and the capsid core containing the HIV-1 genome and proteins essential for infection.

Source: Figure reprinted with permission from (19).

genetic vectors for both experimental and therapeutic purposes. (e) Insertion of retroviral DNA into host chromosomes can be used to mark cell lineages and to make developmental mutants. Progress in these and other areas of retrovirus-related biology has been enormous during the past two decades, but many practical and theoretical problems remain to be solved (20).

2.7 PARAMYXOVIRUSES

Paramyxoviruses are enveloped, single-stranded negative-sense RNA viruses that replicate in the cytoplasm. Diseases caused by these viruses continue to produce high mortality and morbidity across the world. With the development and use of vaccinations and medications, the incidence of serious illness due to paramyxoviruses has tremendously decreased. Yet despite the availability, given the freedom and choice of receiving pre- and/or post-exposure treatment, the cases have increased even in developed countries. Cultivating effective vaccines is still in progress for some of the paramyxoviridae species.

These family of viruses have been recently reclassified. Previously, the *Paramyxoviridae* family of viruses included subfamilies *Paramyxovirinae* and *Pneumovirinae*. However in 2016, the International Committee on the

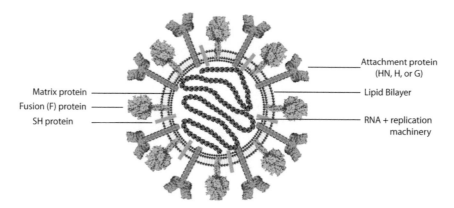

Figure 2.8 Structure of paramyxovirus. Genomic RNA is wrapped by nucleocapsid core proteins (*brown*), which are connected to the viral envelope (*red*) by the matrix protein (*blue*). The attachment (*green*), small hydrophobic (present only in certain paramyxoviruses, *orange*), and fusion proteins (*cyan*) are depicted at the virus surface.

Source: Figure adapted and reprinted with permission from (21).

Taxonomy of Viruses recommended that the paramyxoviruses and pneumoviruses be split into distinct families (*Paramyxoviridae* and *Pneumoviridae*). The reclassification was made for several reasons. First, the polymerase genes of pneumoviruses are more closely related to those of filoviruses than those of paramyxoviruses (Figure 2.8). Second, pneumoviruses differ from paramyxoviruses by possession of an M2 gene that encodes two unique proteins. Third, the ribonucleoprotein (RNP) complexes of pneumoviruses and paramyxoviruses are structurally distinct.

2.8 *ORTHOMYXOVIRIDAE*

These viruses are enveloped, single-stranded negative-sense RNA viruses that replicate in the cytoplasm. The *Orthomyxoviridae* family contains three genera of influenza viruses (influenza A, B, and C viruses) that are classified according to antigenic differences between their nucleoprotein (NP) and matrix 1 (M1) proteins. Such a classification is supported by low intergenic (20%–30%) and high intragenic (>85%) homology of their M1 and NP proteins. Influenza viruses are spherical or filamentous enveloped particles 80 to 120 nm in diameter. The helically symmetric nucleocapsid consists of a nucleoprotein and a multipartite genome of single-stranded antisense RNA in seven or eight segments (Figure 2.9). The envelope carries a hemagglutinin attachment protein and a neuraminidase. The orthomyxoviruses (influenza viruses) constitute the genus *Orthomyxovirus*, which consists of three types

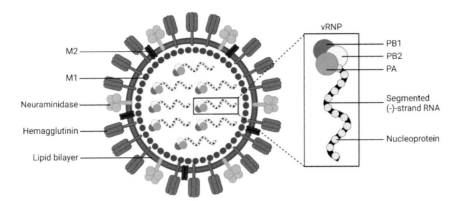

Figure 2.9 Diagram of influenza virus. The enveloped virus contains eight single-stranded RNA segments encoding for each viral protein and are bound to nucleoproteins. Together with the viral polymerase they form the viral ribonucleoprotein (vRNP). The viral RNA-dependent RNA polymerase is a heterotrimer consisting of the polymerase acid (PA), polymerase basic 1 protein (PB1), and polymerase basic 2 protein (PB2). The surface glycoproteins hemagglutinin (HA) and neuraminidase (NA) mediate viral entry and viral release, respectively. The matrix protein (M1) forms a coat inside the virus envelope. The membrane protein (M2) is a proton ion channel, which mediates the acidification of the virus entrapped in the host cell endosome.

Source: Figure reprinted with permission from (22).

(species): A, B, and C. These viruses cause influenza, an acute respiratory disease with prominent systemic symptoms. Pneumonia may develop as a complication and may be fatal, particularly in elderly persons with underlying chronic disease. Type A viruses cause periodic worldwide epidemics (pandemics); both types A and B cause recurring regional and local epidemics. Influenza epidemics have been recorded throughout history (23).

2.9 ARENAVIRIDAE

These are bisegmented ambisense RNA viruses that are a member of the family *Arenaviridae* (Figure 2.10). These viruses infect rodents and occasionally humans. A class of novel, highly divergent arenaviruses, properly known as reptarenaviruses, have also been discovered that infect snakes to produce inclusion body disease. Arenaviruses cause diseases with two types of clinical presentations: neurological and hemorrhagic fever. However, asymptomatic arenavirus infection may occur. Neurological infections include aseptic meningitis, encephalitis, or meningoencephalitis, caused by the LCM virus. The family *Arenaviridae* currently comprises over 20 viral species; several arenaviruses can cause hemorrhagic fever (HF) disease in humans and pose

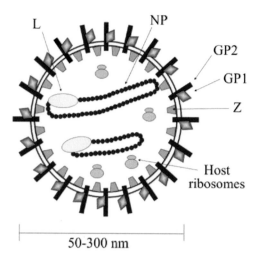

L NP

GP2

GP1

Z

Host ribosomes

50-300 nm

Figure 2.10 Structure of viruses from the Arenavirus family. The mature viral glycoproteins GP1 (*light blue*) and GP2 (*dark blue*) are embedded in the viral envelope (*red*). Beneath the viral envelope is the matrix protein Z (*green*). The ribonucleoprotein complexes are composed of viral RNA encapsidated by the nucleoprotein NP (*purple*) and associated with the RNA-dependent RNA polymerase L (*yellow*).

Source: Figure adapted and reprinted with permission from (24).

a serious public health problem in their endemic regions. Arenaviruses are enveloped viruses with a bisegmented negative-stranded (NS) RNA genome and a life cycle restricted to the cell cytoplasm (25). Virions appear pleomorphic when examined by cryoelectron microscopy, ranging in size from 40 nm to more than 200 nm in diameter. Both genomic viral RNA segments, L (approx. 7.3 kb) and S (approx. 3.5 kb), use an ambisense coding strategy to direct the synthesis of two polypeptides in opposite orientation, separated by a non-coding intergenic region (IGR) that has been shown to act as a bona fide transcription termination signal. There are significant differences in sequence and predicted folded structure between the S and L IGR, but among strains of the same arenavirus species the S and L IGR sequences are highly conserved. The S RNA encodes the viral NP and the glycoprotein precursor (GPC) that is post-translationally cleaved by the cellular site 1 protease (S1P) to yield the two mature virion glycoproteins GP1 and GP2 that form the spikes that decorate the virus surface. GP1 mediates virus receptor recognition and cell entry via endocytosis, whereas GP2 mediates the pH-dependent fusion event required to release the virus ribonucleoprotein (RNP) core into the cytoplasm of infected cells. The L RNA encodes the viral RNA-dependent RNA polymerase (RdRp, or L polymerase) and the small RING finger protein Z that has functions of a bona fide matrix protein including directing virus budding (26).

2.10 *RHABDOVIRIDAE*

Rhabdoviridae consists of more than 100 single-stranded, negative-sense, non-segmented viruses that infect a wide variety of hosts, including vertebrates, invertebrates, and plants. Common to all members of the family is a distinctive rod- or bullet-shaped morphology. Human pathogens of medical importance are found in the genera *Lyssavirus* and *Vesiculovirus*. Rabies virus is medically the most significant member of the genus *Lyssavirus*. Rhabdoviruses, as single-stranded, negative-sense RNA viruses within the order *Mononegavirales*, are characterized by bullet-shaped or bacteroid particles that contain a helical ribonucleoprotein complex (RNP) (27). Rabies virus causes acute infection of the CNS. It is a rod- or bullet-shaped, single-stranded, negative-sense, unsegmented, enveloped RNA virus (Figure 2.11). The virus genome

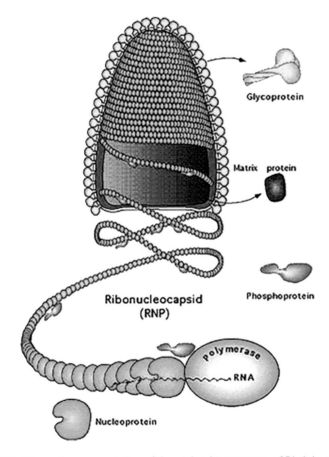

Figure 2.11 Schematic representation of the molecular structure of Rhabdovirus.

Source: Figure reprinted with permission from (28).

encodes five proteins. After inoculation, rabies virus may enter the peripheral nervous system directly and migrate to the brain or may replicate in muscle tissue, remaining sequestered at or near the entry site during incubation, prior to CNS invasion and replication. It then spreads centrifugally to numerous other organs. Five general stages of rabies are recognized in humans: incubation, prodrome, acute neurologic period, coma, and death (or, very rarely, recovery). No specific antirabies agents are useful once clinical signs or symptoms develop. The incubation period in rabies, usually 30 to 90 days but ranging from as few as 5 days to longer than 2 years after initial exposure, is more variable than in any other acute infection. Incubation periods may be somewhat shorter in children and in individuals bitten close to the CNS (e.g., the head). Clinical symptoms are first noted during the prodromal period, which usually lasts from 2 to 10 days. These symptoms are often nonspecific (general malaise, fever, and fatigue) or suggest involvement of the respiratory system (sore throat, cough, and dyspnea), gastrointestinal system (anorexia, dysphagia, nausea, vomiting, abdominal pain, and diarrhea), or CNS (headache, vertigo, anxiety, apprehension, irritability, and nervousness). More remarkable abnormalities (agitation, photophobia, priapism, increased libido, insomnia, nightmares, and depression) may also occur, suggesting encephalitis, psychiatric disturbances, or brain conditions. Pain or paresthesia at the site of virus inoculation, combined with a history of recent animal bite, should suggest a consideration of rabies.

2.11 *FILOVIRIDAE*

In this family viruses are enveloped with a negative-strand RNA noted for their ability to cause outbreaks of severe, often fatal, viral hemorrhagic fevers. The virus family *Filoviridae* is divided into three genera: *Ebolavirus*, *Marburgvirus*, and *Cuevavirus*. Filoviruses possess single-stranded, negative-sense RNA genomes of approximately 19 kilobases. They have seven distinct genes, each of which functions as a separate transcriptional unit (F). These encode seven viral structural proteins known as nucleoprotein (NP), viral protein (VP) 35, VP40, glycoprotein (GP), VP30, VP24, and the large protein (L). For the Ebolaviruses, the GP gene encodes multiple different proteins due to transcriptional editing by the viral RNA-dependent RNA polymerase (RDRP) as shown in Figure 2.12. The non-template encoded "As" in Ebola virus are added co-transcriptionally to the nascent mRNA. The unedited transcript encodes sGP, while editing results in translational frameshifting such that the membrane-bound GP and a second secreted GP are produced from edited RNAs (29). Additional editing events recently described for both EBOV and MARV mRNAs may further diversify the proteins translated from filovirus mRNAs. Human infections are acquired by direct contact with infected bodily fluid and likely enter the human body via breaks in the skin or through mucosal surfaces. Studies using nonhuman primates,

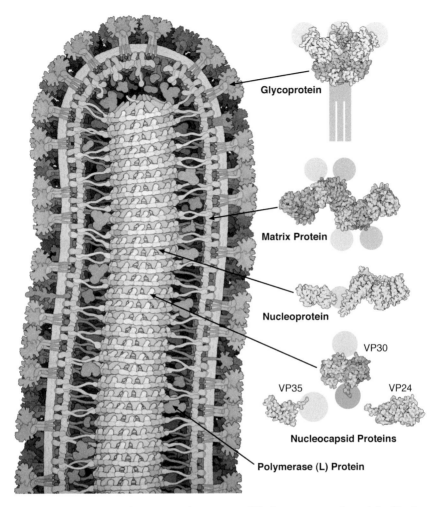

Figure 2.12 Molecular architecture and structure of Ebola virus, a member of the Filovirus family of RNA viruses.

Source: Figure adapted and reprinted with permission from (30).

the gold standard animal model for studying EBOV disease, have provided significant insight into the pathogenesis of filovirus infections (31).

2.12 BUNYAVIRIDAE

Bunyaviruses are spherical, enveloped particles 90 to 100 nm in diameter. They contain single-stranded RNA, which, with the nucleoprotein, forms three nucleocapsid segments. The segments are large, medium, and small

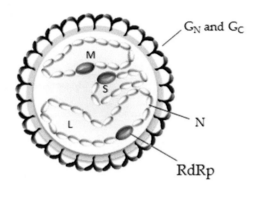

G_N and G_C

RdRp

80–160 nm

Figure 2.13 Structure of Bunyavirus. The three viral genomic segments are termed according to their size: S (small), M (medium) and L (large). The largest genomic RNA segment is the (L) fragment that encodes the RNA-dependent RNA polymerase L, an enzyme that is critical for viral replication initiation after the virus genome is released into the cytoplasm of host cells. The medium virus RNA segment (M) encodes a polypeptide precursor that gets processed in the endoplasmic reticulum or Golgi apparatus into two envelope glycoproteins, G_N and G_C. G_N, glycoprotein G_N; G_C, glycoprotein G_C; N, nucleoprotein; RdRp, RNA-dependent RNA polymerase.

Source: Figure reprinted with permission from (32).

helical, circular structures. The RNA has a total molecular weight of 5×10^6 daltons. The nucleocapsid is surrounded by a lipid-containing envelope. Surface spikes are composed of two glycoproteins that confer properties of neutralization of infectivity and hemagglutination of red blood cells. *Bunyaviridae* is a family of arthropod-borne or rodent-borne, spherical, enveloped RNA viruses (Figure 2.13). Bunyaviruses are responsible for several febrile diseases in humans and other vertebrates. They have either a rodent host or an arthropod vector and a vertebrate host. The *Bunyaviridae* are divided into arthropod-borne viruses (arboviruses) and rodent-borne viruses (roboviruses). Bunyaviruses cause several diseases of human and domestic animals, including fever, hemorrhagic fever, renal failure, encephalitis, meningitis, blindness, and, in domestic animals, congenital defects. Most illnesses are self-limited fevers that last 1 to 4 days and are accompanied by headache, muscle aches, nausea, conjunctival injection, and generalized weakness. A few are more serious illnesses: La Crosse encephalitis is characterized by fever, convulsions, drowsiness, and focal neurologic signs; Crimean-Congo hemorrhagic fever is characterized by headache, pain in limbs, and, in severe cases, bleeding from multiple orifices; hemorrhagic fever with renal syndrome (Korean hemorrhagic fever, nephropathia epidemica) is characterized by fever, hemorrhage, and acute renal failure; and hantavirus pulmonary syndrome is characterized by fever and acute respiratory distress. Rift Valley fever may mimic the febrile, encephalitic, or hemorrhagic illness of other bunyavirus infections, and the patient may also go blind because of retinal vasculitis. These illnesses are significant, currently uncontrolled human diseases. La Crosse virus causes most of the arbovirus encephalitis in North America. Also, more than 100,000 cases of hemorrhagic fever with renal syndrome occur annually in Asia and Europe. Rift Valley fever has explosive potential, as shown in Egypt in 1977, when an estimated 200,000 cases, with 598 deaths, were recorded. Hantavirus pulmonary syndrome is uncommon but is associated with a 50% case fatality rate. Bunyaviruses cause fevers sometimes with rash. In addition, Crimean-Congo hemorrhagic fever virus may cause hemorrhage; Rift Valley fever virus may cause hemorrhagic hepatitis, encephalitis, or blindness; La Crosse virus and related viruses may cause encephalitis; and Hantavirus and related viruses may cause hemorrhage and renal failure or the hantavirus pulmonary syndrome (33).

2.13 CONCLUSION

Viral hemorrhagic fevers (VHFs) are a group of diseases that are caused by several distinct families of viruses. The term "viral hemorrhagic fever" refers to a condition that affects many organ systems of the body, damages the overall cardiovascular system, and reduces the body's ability to function on its own. Symptoms of this type of condition can vary but often include bleeding or hemorrhaging. Some VHFs cause relatively mild illness, while others can cause severe, life-threatening disease. Most VHFs have no known cure or vaccine.

REFERENCES

1. Carrasco-Hernandez R, Jácome R, López Vidal Y, Ponce de León S. Are RNA viruses candidate agents for the next global pandemic? A review. *ILAR Journal* [Internet] 2017;58(3); 343–358. Available from: https://academic.oup.com/ilarjournal/article/58/3/343/4107390
2. Gelderblom HR. Structure and classification of viruses [Internet]. *Medical Microbiology* 1996. Available from: www.ncbi.nlm.nih.gov/pubmed/142303
3. Smertina E, Urakova N, Strive T, Frese M. Calicivirus RNA-Dependent RNA polymerases: Evolution, structure, protein dynamics, and function. *Frontiers in Microbiology* [Internet] 2019; 10. Available from: www.frontiersin.org/article/10.3389/fmicb.2019.01280/full
4. Campillay-Véliz CP, Carvajal JJ, Avellaneda AM, Escobar D, Covián C, Kalergis AM, et al. Human norovirus proteins: Implications in the replicative cycle, pathogenesis, and the host immune response. *Front Immunology* [Internet] 2020; 11. Available from: www.frontiersin.org/article/10.3389/fimmu.2020.00961/full
5. Liu D. No title. In *MDPI* [Internet]. MDPI: Basel; 2022. Available from: https://encyclopedia.pub/entry/27045

6. Martinez-Salas E, Fernandez-Miragall O. Picornavirus IRES: Structure function relationship. *Current Pharmaceutical Design* [Internet] 2004;10(30); 3757–3767. Available from: www.eurekaselect.com/openurl/content.php?genre=article&issn=1381-6128&volume=10&issue=30&spage=3757

7. Martínez-Salas E, Francisco-Velilla R, Fernandez-Chamorro J, Lozano G, Diaz-Toledano R. Picornavirus IRES elements: RNA structure and host protein interactions. *Virus Research* [Internet] 2015;206; 62–73. Available from: www.ncbi.nlm.nih.gov/pubmed/25617758

8. Go YY, Balasuriya UBR, Lee C. Zoonotic encephalitides caused by arboviruses: Transmission and epidemiology of alphaviruses and flaviviruses. *Clinical and Experimental Vaccine Research* [Internet] 2014;3(1); 58. Available from: https://ecevr.org/DOIx.php?id=10.7774/cevr.2014.3.1.58

9. Lang L. The food and drug administration approves second West Nile virus screening test for donated blood and organs. *Gastroenterology* [Internet] 2007;133(5); 1402. Available from: https://linkinghub.elsevier.com/retrieve/pii/S0016508507016873

10. Souza LR, Colonna JG, Comodaro JM, Naveca FG. Using amino acids co-occurrence matrices and explainability model to investigate patterns in dengue virus proteins. *BMC Bioinformatics* [Internet] 2022;23(1); 80. Available from: https://bmcbioinformatics.biomedcentral.com/articles/10.1186/s12859-022-04597-y

11. Maghoul A, Simonsen I, Rostami A, Mirtaheri P. An optical modeling framework for coronavirus detection using graphene-based nanosensor. *Nanomaterials* [Internet] 2022;12(16); 2868. Available from: www.mdpi.com/2079-4991/12/16/2868

12. Chen HX, Chen ZH, Shen HH. [Structure of SARS-CoV-2 and treatment of COVID-19]. *Sheng Li Xue Bao* [Internet] 2020;72(5); 617–630. Available from: www.ncbi.nlm.nih.gov/pubmed/33106832

13. Dash S, Dash C, Pandhare J. Therapeutic significance of microRNA-mediated regulation of PARP-1 in SARS-CoV-2 infection. *Non-Coding RNA* [Internet] 2021;7(4); 60. Available from: www.mdpi.com/2311-553X/7/4/60

14. Johnson RT. Retroviruses and nervous system disease. *Current Opinion in Neurobiology* [Internet] 1992;2(5); 663–670. Available from: https://linkinghub.elsevier.com/retrieve/pii/095943889290036K

15. Toledano M, Aksamit AJ. Retroviral infections of the nervous system. In *Mayo Clinic Neurology Board Review* [Internet]. Oxford: Oxford University Press; 2015, 611–616. Available from: https://academic.oup.com/book/42613/chapter/357620825

16. Spudich S, Gonzalez-Scarano F. HIV-1-Related central nervous system disease: Current issues in pathogenesis, diagnosis, and treatment. *Cold Spring Harbor Perspectives in Biology* [Internet] 2012;2(6); a007120–a007120. Available from: http://perspectivesinmedicine.cshlp.org/lookup/doi/10.1101/cshperspect.a007120

17. Dash S, Balasubramaniam M, Villalta F, Dash C, Pandhare J. Impact of cocaine abuse on HIV pathogenesis. *Frontiers in Microbiology* [Internet] 2015; 6. Available from: http://journal.frontiersin.org/Article/10.3389/fmicb.2015.01111/abstract

18. Pandhare J, Dash S, Jones B, Villalta F, Dash C. A novel role of proline oxidase in HIV-1 envelope glycoprotein-induced neuronal autophagy. *Journal of Biological Chemistry* [Internet] 2015;290(42); 25439–25451. Available from: https://linkinghub.elsevier.com/retrieve/pii/S0021925820445624

19. Rossi E, Meuser ME, Cunanan CJ, Cocklin S. Structure, function, and interactions of the HIV-1 capsid protein. *Life* [Internet] 2021;11(2); 100. Available from: www.mdpi.com/2075-1729/11/2/100

20. Varmus H. Retroviruses. *Science (80-)* [Internet] 1988;240(4858); 1427–1435. Available from: www.science.org/doi/10.1126/science.3287617

21. Chang A, Dutch RE. Paramyxovirus fusion and entry: Multiple paths to a common end. *Viruses* [Internet] 2012;4(4); 613–636. Available from: www.mdpi.com/1999-4915/4/4/613

22. Eichberg J, Maiworm E, Oberpaul M, Czudai-Matwich V, Lüddecke T, Vilcinskas A, et al. Antiviral potential of natural resources against influenza virus infections. *Viruses* [Internet] 2022;14(11); 2452. Available from: www.mdpi.com/1999-4915/14/11/2452

23. Hause BM, Collin EA, Liu R, Huang B, Sheng Z, Lu W, et al. Characterization of a novel influenza virus in cattle and swine: Proposal for a new genus in the Orthomyxoviridae family. *MBio* [Internet] 2014;5(2); e00031–e00014. Available from: www.ncbi.nlm.nih.gov/pubmed/24595369

24. Wolff S, Ebihara H, Groseth A. Arenavirus budding: A common pathway with mechanistic differences. *Viruses* [Internet] 2013;5(2); 528–549. Available from: www.mdpi.com/1999-4915/5/2/528

25. Grande-Pérez A, Martin V, Moreno H, de la Torre JC. *Arenavirus Quasispecies and Their Biological Implications*; 2015, 231–75. Available from: http://link.springer.com/10.1007/82_2015_468

26. Strecker T, Eichler R, Meulen JT, Weissenhorn W, Dieter Klenk H, Garten W, et al. Lassa virus Z protein is a matrix protein sufficient for the release of virus-like particles. *Journal of Virology* [Internet] 2003;77(19); 10700–107005. Available from: https://journals.asm.org/doi/10.1128/JVI.77.19.10700-10705.2003

27. Riedel C, Hennrich AA, Conzelmann K-K. Components and architecture of the rhabdovirus ribonucleoprotein complex. *Viruses* [Internet] 2020;12(9); 959. Available from: www.mdpi.com/1999-4915/12/9/959

28. Rhabdoviridae. In *Virus Taxonomy* [Internet]. Elsevier; 2012, 686–713. Available from: https://linkinghub.elsevier.com/retrieve/pii/B9780123846846000574

29. Negredo A, Palacios G, Vázquez-Morón S, González F, Dopazo H, Molero F, et al. Discovery of an Ebolavirus-like filovirus in Europe. In Basler CF, ed., *PLoS Pathogens* [Internet] 2011;7(10); e1002304. Available from: https://dx.plos.org/10.1371/journal.ppat.1002304

30. Feldmann H. Ebola—A growing threat? *New England Journal of Medicine* [Internet] 2014;371(15); 1375–1378. Available from: www.nejm.org/doi/10.1056/NEJMp1405314

31. Messaoudi I, Amarasinghe GK, Basler CF. Filovirus pathogenesis and immune evasion: Insights from Ebola virus and Marburg virus. *Nature Reviews Microbiology* [Internet] 2015;13(11); 663–676. Available from: www.nature.com/articles/nrmicro3524

32. Albornoz A, Hoffmann A, Lozach PY, Tischler N. Early bunyavirus-host cell interactions. *Viruses* [Internet] 2016;8(5); 143. Available from: www.mdpi.com/1999-4915/8/5/143

33. Shope RE. Bunyaviruses [Internet]. *Medical Microbiology* 1996. Available from: www.ncbi.nlm.nih.gov/pubmed/7811854

Chapter 3

Common Neuropathological Features of RNA Viruses

Anne Khodarkovskaya

3.1 ENCEPHALOPATHY AND ENCEPHALITIS

Encephalopathy is a broad term used to describe disease or damage in the brain that results in abnormal brain structure and/or function. Symptomatically, it presents itself as confusion or a change in mental state, affecting personality, memory, and cognitive abilities. Additional symptoms can include seizures, twitching (**myoclonus**), rapid involuntary eye movement (**nystagmus**), loss of speech or comprehension (**aphasia**), hand tremor (**asterixis**), muscle atrophy, fatigue, altered consciousness, breathing or swallowing difficulties, and other related symptoms (1). Encephalopathy can emerge from a wide range of causes, such as traumatic injury, metabolic and systemic disease, exposure to toxins, lack of oxygen (**hypoxia** or **anoxia**), tumor or increased intracranial pressure, and infectious agents like RNA viruses. Due to the non-specific nature of this condition, magnetic resonance imaging is a valuable tool in early diagnosis and detection (2,3) (Figure 3.1).

Encephalitis refers to a specific form of encephalopathy consisting of inflammation or swelling in the brain and parenchymal central nervous system (CNS) tissues. While some cases do not lead to symptoms, symptoms can range from mild and flulike, such as fever and headache, to severe presentations like delirium, seizures, personality changes, aphasia, impaired coordination (**ataxia**), and muscle weakness in limbs (**quadriparesis**) (4). Stiffness in the neck can indicate inflamed meninges in addition to parenchymal inflammation, which is described as **meningoencephalitis**. If the spinal cord also exhibits inflammation with damage to myelin, it is referred to as **encephalomyelitis**. Encephalitis localized to certain areas of the brain can have specific symptoms, such as disorientation or memory loss, as in the cases of **rhombencephalitis** or **limbic encephalitis** (5). Lumbar puncture in cases of encephalitis can show mildly elevated white blood cell count (**pleocytosis**) and moderate elevation of cerebrospinal fluid (CSF) proteins (5,6).

Encephalitis can result **secondary** to viral infections, although some viruses cause **primary** encephalitis through direct invasion of the CNS. Primary encephalitis typically shows neuronal loss and tissue necrosis in the gray matter (neurotropism), releasing proinflammatory cytokines as a result, while secondary encephalitis appears more in white matter through inflammation and demyelination and can also follow primary encephalitis (7,8). Viral infections can also lead to **autoimmune encephalitis**, in which the body's immune system mistakenly targets healthy neuronal and glial

DOI: 10.1201/9781003285823-3

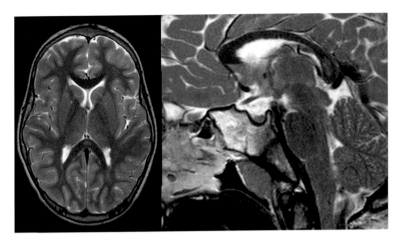

Figure 3.1 Mild viral encephalitis with a reversible lesion in the splenium (posterior corpus callosum) indicated by T2 hyperintensity on MRI in transverse (*left*) and sagittal (*right*) sections of the brain.

Source: Reprinted with permission from (9).

tissue, which can lead to inflammation (3,10). Autoimmune encephalitis can range from antibodies against voltage-gated potassium channels, resulting in memory loss, confusion, behavioral changes, and severe degeneration in the hippocampus, to *N*-methyl-D-aspartate (NMDA) receptor–associated encephalitis, which shows little inflammation or neuronal loss but can lead to seizures, neuropsychological disturbances, dyskinesias, loss of consciousness, and autonomic dysregulation (8).

More than 100 viruses can cause encephalitis and encephalopathies. Among RNA viruses that can lead to acute encephalopathies are influenza, rotavirus, human respiratory syncytial virus, coronavirus, human metapneumovirus, and others. RNA viruses that can lead to encephalitis include rabies virus, measles, mumps, Japanese encephalitis virus, tick-borne encephalitis virus, West Nile virus, SARS-CoV-2, Zika virus, dengue virus, St. Louis encephalitis virus, Eastern and Western equine encephalitis viruses, poliovirus, and more (4,11). RNA viruses can lead to encephalopathies that arise due to a number of reasons including hypoxia or liver disease. Encephalopathy that results from infectious agents is typically referred to as **toxic-metabolic encephalopathy**. Toxic-metabolic encephalopathy can result from endocrine and metabolic imbalances (e.g., in sodium, calcium, or glucose), liver dysfunction, renal dysfunction, sepsis, cytokine storm, and inflammation (12). While this form of encephalopathy tends to be acute, features of viral encephalopathies can include acute necrotizing encephalopathy, classical Reye syndrome, Reye-like syndrome, infantile (frontal lobe) encephalopathy, and others. Reye syndrome is rare but can occur after viral infection from influenza, leading to both brain and liver swelling as a result of mito-chondrial and metabolic dysfunction. According to Morichi et al. (2012) and building on Mizuguchi et al. (2007), acute encephalopathies caused by viruses such as influenza and human respiratory syncytial virus (hRSV) can be classified into four subtypes:

1. Metabolic error
2. Cytokine storm
3. Excitotoxicity, characterized by convulsions, seizures, and eventually coma
4. Hypoxic encephalopathy (13,14)

The incidence of acute encephalopathies is highest among infants and children. Influenza, a negative-sense, single-stranded RNA virus, is one of the biggest players with regards to the incidence of acute viral encephalopathy, accounting for 40%–50% of acute necrotizing encephalopathies and Reye-like syndrome (14). Encephalitis and encephalopathy constitute 37.7% of influenza-associated neurologic complications (15). Influenza virus that does not appear in the CSF of infected patients does not seem to directly invade the CNS, meaning encepha-litis and encephalopathy result as post-infection sequelae likely due to cytokine storm (5). Cytokine storm following severe infection can lead to acute encepha-lopathy due to widespread organ failure and inability of the liver to process toxins and produce toxic metabolites. Inflammatory cytokines in the brain may lead to further excitotoxicity and neuronal death (14).

hRSV is a negative-sense, single-stranded RNA virus that can infect peo-ple of all ages but is a highly contagious, common childhood illness that most children under the age of two contract. It typically produces mild cold symptoms but can cause severe infection in infants or elderly adults with weakened immune systems. In these cases, acute encephalopathy can occur in a small percentage of patients (about 1%–2%), which occasionally leads to seizures or convulsions (16). Of the subtypes previously mentioned, the excitotoxicity type is the most common. Hypoxic encephalopathy is a pos-sible sequela of hRSV, influenza, and other respiratory viruses (13).

Positive-sense, single-stranded, enveloped RNA flaviviruses, such as Japa-nese encephalitis virus, tick-borne encephalitis virus, West Nile virus, dengue virus, St. Louis encephalitis virus, and Zika virus, are neurotropic viruses that directly infect cells in the brain and lead to encephalitis. Viruses can enter the CNS through the olfactory route, crossing the blood–brain barrier (BBB) and causing neuronal cell death. Additionally, they can propagate in astrocytes, which are particularly resistant to apoptosis from flaviviruses (17). Increased viral production in astrocytes leads to upregulation of proinflammatory cyto-kines and BBB breakdown. The viruses can further infect microglia, pericytes, and brain microvascular endothelial cells, further disrupting the tight junc-tions of the BBB and promote secondary neuronal damage (18).

Rabies virus, a negative-sense, single-stranded, enveloped RNA virus, is another example of a primary encephalitis-causing virus. Once the virus enters muscle, it can spread to the peripheral nervous system through neuromuscular junctions and travel to the CNS via retrograde axonal transport. Increased production of nitric oxide and reactive oxygen species by neurons and macrophages results in mitochondrial dysfunction followed by axonal swelling. Symptoms manifest after the virus reaches the brainstem, resulting in **hydrophobia** and **aerophobia** (due to difficulty swallowing and breathing) in 50%–80% of patients, and severe cases can lead to myoclonus, paralysis, coma, and death. In certain hosts, neuronal apoptosis or loss of axons and dendrites can occur to inhibit further spread of the virus (19).

3.2 MENINGITIS AND MYELITIS

Meningitis is an inflammation of the meninges, the membranes surrounding the brain and spinal cord. Its key symptom is neck stiffness, demonstrable by Brudzinski's sign and Kernig's sign. In severe cases, Brudzinski's sign shows a flexed, stiff neck causing a patient's hips and knees to flex. Kernig's sign results in stiffness in the hamstring such that the leg cannot be straightened when the hip is flexed at 90 degrees. Additional symptoms include fever, headache, an aversion to bright lights (**photophobia**) and loud noises (**phonophobia**), drowsiness, nausea, rash, and seizures. Meningitis is most often secondary to infectious agents that lead to edema and inflammation in the CNS and is referred to as **aseptic meningitis** when caused by viruses, as opposed to bacteria. It can be diagnostically confirmed by lumbar puncture showing elevated mononuclear cells in the CSF (**pleocytosis**). RNA viruses that can lead to meningitis include mumps, measles, several flaviviruses, and the enteroviruses which account for 80% of viral meningitis cases today (20).

Meningitis is a common symptom of mumps, occurring in about 15% of cases. Prior to widespread vaccination, it was the leading cause of meningitis in young children. While mumps usually presents with parotid swelling and fever, meningitis can still arise in the absence of these symptoms, preceding or following parotitis. CSF samples show that the virus involves the CNS in at least half of all cases, with the virus being detectable in CSF by reverse transcription–polymerase chain reaction in most cases up to 2 years later. Interestingly, young men seem to be at greater risk for encephalomyelitis. In rare cases, mumps can lead to permanent damage from myelitis (21).

Myelitis is an inflammation of the spinal cord. It can disrupt signals between the brain and the rest of the body and can result in permanent damage to the myelin and axons of the spinal cord. Symptoms can include loss of sensation (**paresthesia**) or muscle weakness and paralysis. Myelitis concomitant with meningitis can be referred to as **meningomyelitis, or meningococcal meningitis**. It can be a result of direct infiltration of the CNS by infectious agents or the result of an autoimmune condition. The infectious agents can include RNA viruses like enteroviruses, flaviviruses, influenza, and HIV (22). A lumbar puncture can be performed to diagnose myelitis based on elevated proteins in the CSF (7).

Enteroviruses like poliovirus, a single-stranded, positive-sense, nonenveloped virus, can cause both meningitis and myelitis. Around 1% of cases lead to aseptic meningitis, characterized by spasms along the neck, back, and limbs (23). In a smaller percentage of cases that progress past flulike symptoms, polioviruses can infect and replicate in the CNS and result in neurodegeneration and subsequent paralysis. This poliomyelitis is characterized by motor neuron loss in the anterior horn of the spinal cord (24,25) (Figure 3.2). Damage to the anterior horn results in limb paralysis. The virus

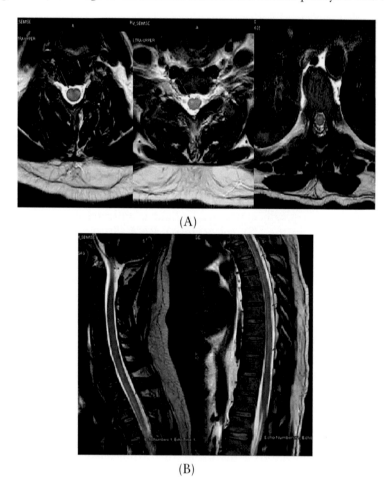

(A)

(B)

Figure 3.2 MRI of acute transverse myelitis associated with SARS-CoV-2 infection. T2 hyperintensities are visible in the anterior horn of the cervicothoracic spinal cord in (A) transverse section and (B) sagittal section.

Source: Reprinted with permission from (26).

can spread further, including the posterior horns, the motor neurons of the thalamus, and hypothalamus. Spreading to the brainstem can interfere with the ability to breathe and result in fatality. Muscle atrophy can continue 25 to 30 years after first contracting polio in post-polio syndrome, possibly due to continued neuronal degeneration or persistence of the virus causing abnormal presence of cytokines (23).

3.3 NEURODEGENERATION

Neurodegenerative diseases (NDs) are disorders of the nervous system marked by the loss of neuronal structure and function. They can lead to cognitive, sensory, and motor dysfunctions and are the fourth leading cause of death. The etiology and cause for several NDs are unknown and seem to arise from several possible avenues. The incidence of NDs increases with age, pointing to aging-related inflammation as a cause; however, there is documented evidence that several viruses can influence the onset of NDs (27). Viruses are able to directly injure neurons, induce autoimmune processes, or induce cellular changes or cascades that resemble NDs (28,29). Among these NDs are multiple sclerosis, Parkinson's disease (PD), Guillain–Barré syndrome, amyotrophic lateral sclerosis (ALS), and dementia. RNA viruses that have been implicated in these disorders include coronaviruses, influenza viruses, Zika virus, and retroviruses such as HIV (27,28).

Multiple sclerosis (MS) is a chronic autoimmune demyelinating condition of the CNS, characterized by focal white and gray matter lesions (30). The clinical symptoms can manifest as muscle weakness and paralysis, pain, muscle spasms, and loss of sensation (28). Coronaviruses, which are positive-sense, single-stranded enveloped viruses, have long been implicated as a possible cause of MS (5). In fact, murine coronaviruses like mouse hepatitis virus are often used to produce mouse models of MS, as the virus produces a similar demyelination pattern and encephalomyelitis (31–34). It is hypothesized that coronaviruses can lead to autoimmune demyelination because many share epitopes with a string of four to six peptides similar to myelin basic protein (MBP), causing the host to induce immunity against MBP (35). Additionally, coronaviruses have been isolated from CNS samples of human MS patients (31,33). Autopsy samples of several neurological diseases, with MS being most prominent, showed presence of coronaviruses in 67% of samples (4). Coronaviruses have been shown to infect neuronal and glial cells, including microglia, macrophages, astrocytes, and brain microvessel endothelial cells (4,35). Loss of trophic support from glial cells can lead to axonal atrophy (34). Further neuronal damage can result from coronaviruses by way of hypoxia, due to the respiratory and inflammatory implications of the disease (33).

Coronaviruses also may be relevant in Parkinson's disease. PD is characterized by loss of dopamine neurons in the substantia nigra, as well as aggregation of α-synuclein, symptomatically presenting with bradykinesia, rigidity, and tremor. During aging and chronic inflammation, the substantia nigra becomes more susceptible to environmental influences. Cells of the substantia nigra are enriched with SARS-CoV-2 receptors ACE2 and TMPRSS2 and may be susceptible to infection (36). These cells could be sensitive to mitochondrial injury and oxidative stress. SARS-CoV-2 may also induce microglial dysfunction leading to inflammation, protein aggregation, neurodegeneration, and autophagy dysfunction. Anecdotally, severe infection with SARS-CoV-2 has been correlated to the onset or worsening of PD (37). Other RNA viruses previously implicated in Parkinson's include hepatitis C virus, HIV, Japanese encephalitis virus, West Nile virus, and influenza A. Both influenza and SARS-CoV have been shown to modulate cellular aging pathways that can lead to PD (36).

Correlation between influenza A and PD has been investigated since the early 20th century. After the influenza A pandemic, an outbreak of encephalitis lethargica occurred, followed by postencephalitic parkinsonism. People born during the pandemic displayed a two- to three-fold risk of developing parkinsonism compared to people born before or after (36). Postmortem studies suggested a causative effect, with loss of dopaminergic neurons in the substantia nigra, neurofibrillary tangles, and presence of influenza in the dopamine neurons (27,38) (Figure 3.3). Similar outcomes have been seen in animal models infected with influenza strains (38). Influenza A virus has been shown to infect dopaminergic neurons of the substantia nigra and lead to α-synuclein aggregation via a mechanism related to inhibition of autophagy and microglial activation (4,27,29,36).

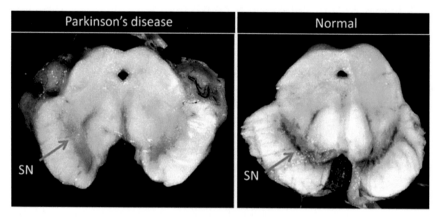

Figure 3.3 Loss of neuromelanin and dopamine neurons in the substantia nigra, a characteristic of Parkinson's disease.

Source: Reprinted with permission from (39).

Influenza can disrupt cellular autophagy in neurons, inducing misfolded proteins that can then contribute to parkinsonism molecular processes (29). Although no direct correlation has been made between influenza and the pathogenesis of PD, it is a possible etiological factor due to its role in neuroinflammation and explaining Parkinson's clusters, where people in close contact with each other develop PD at higher incidences than the general population (27).

The influenza virus is also one of several viruses implicated in **Guillain–Barré syndrome** (GBS). GBS is an acute inflammatory demyelinating condition that can arise after infections, believed to be due to a molecular mimicry mechanism between the infectious agent and peripheral nerves leading to autoimmune disorder (12). It affects less than 0.004% of the general population (40). Anecdotally, some coronavirus cases have led to GBS, but numbers are low and correlation has been debated (3,12,35). Other RNA viruses associated with GBS include Zika virus, dengue virus, chikungunya virus, West Nile virus, and Japanese encephalitis virus (24). GBS constitutes 1.3% of neurologic complications of influenza (5). About 60% of GBS cases have unknown causes but peak in the winter, indicating influenza viruses play a significant role during outbreaks. However, influenza-related cases did not show antiganglioside antibodies, which suggests a separate autoimmune method than others documented (40). Zika virus, though infections are usually mild, can also lead to GBS in adults in an estimated 1.2% of infections. Similarly, typical antiganglioside antibodies were not noted, but antiglycolipid antibodies may be present (24).

3.4 NEUROAIDS

Human immunodeficiency virus (HIV), a retrovirus of two noncovalently linked single-stranded, positive-sense RNA, can lead to neuron loss that manifests in motor and cognitive disorders. HIV-associated cognitive impairment affects more than 30% of people who contract the virus and 60% of the population that develops acquired immune deficiency syndrome (AIDS), despite the availability of antiretroviral therapies because of their limitations in the CNS (27,41). The most detrimental of HIV-associated neurocognitive disorders is **HIV-associated dementia** (HAD). HAD appears symptomatically with decreased attention, concentration, psychomotor speed, learning, memory, processing, executive function, and behavior changes like apathy, and irritability (42). HIV can infect astrocytes, macrophages, and microglia, altering the integrity of the BBB to access the CNS, leading to devastating consequences in the brain, inducing neuronal dysfunction and apoptosis (35). HAD is characterized by perivascular macrophage infiltration, microglial activation and nodules, multinucleated giant cells, reactive astrocytosis, subacute encephalitis, and myelin sheath damage visible by myelin pallor in addition to neuron loss (27,32). HIV can further inhibit amyloid degradation and increase phosphorylation of tau protein, leading to amyloid plaques and neurofibrillary tangles similar to those seen in **Alzheimer's disease** (AD), in addition to other age-related factors of AD. HAD shares some similarities with **Lewy body dementia** and PD, exhibiting increased α-synuclein in the substantia nigra, dopamine deficiency, and testosterone deficiency (27,42).

Neurological complications of HIV and AIDS also include motor and sensory disorders. Symptoms can include motor slowing, incoordination, tremor, weakness, spasticity, extrapyramidal movement disorders, and paraparesis (42). In addition to Parkinson's-like symptoms, connections have been drawn to MS and **amyotrophic lateral sclerosis**. ALS is characterized by neurodegeneration in the spinal cord and cortical neurons and, although the cause is unknown, it has been associated with some viruses, such as HIV and other retroviruses due to presence of reverse transcriptase in ALS patients' sera comparable to that of HIV patients (27) (Figure 3.4). Additionally, ALS is characterized by motor neuron death as a result of TDP-43 and FUS protein aggregation and mislocalization, both of which are proteins that repress

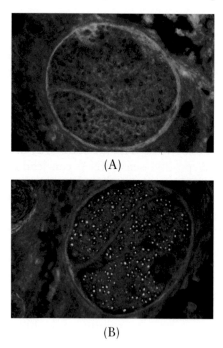

(A)

(B)

Figure 3.4 Human nerve tissue incubated with (A) normal human serum and (B) HIV-positive human serum from a patient with HIV-related neuropathy, showing anti-myelin antibodies.

Source: Reprinted with permission from (43).

HIV transcription. Viral infections can trigger aggregation of these proteins, thus virus-induced multiprotein complexes may lead to neurodegenerative disorders such as ALS (29). Retroviruses in animal models have been shown to induce spongiform change, anterior horn motor neuron loss, reactive gliosis, inflammation, microglial nodules, and cerebral atrophy (32). Furthermore, late manifestations of HIV disease have shown **vacuolar myelopathy** and **peripheral neuropathy**. Most HIV patients with vacuolar myelopathy and sensory neuropathy present with immunosuppression, spastic paraparesis, gait disorders, lower extremity sensory abnormalities, urinary abnormalities, ataxia, impaired proprioception, encephalopathy, and dementia. MRI shows spinal cord atrophy in the posterior and lateral columns. It is thought to be related to impaired myelin metabolism and subsequent neurotoxic cytokines (44).

3.5 ANOSMIA AND AGEUSIA

Anosmia or hyposmia is the loss of smell, and **ageusia** or hypogeusia is the loss of taste. These symptoms are often linked to viral upper respiratory tract infections, as many viruses induce inflammatory reactions that can alter smell. RNA viruses that can cause these symptoms are coronaviruses, influenza, parainfluenza viruses, and rhinoviruses (45). One entrance to the CNS for viruses is through the nose and olfactory bulb via the trigeminal nerve and the olfactory nerve, which was originally thought to be the main mechanism of anosmia in COVID-19 (29,46). Recent data supports the theory that the anosmia is a result of SARS-CoV-2 direct invasion of olfactory epithelium and vascular pericytes, while olfactory sensory and bulb neurons remain intact (47).

SARS-CoV-2, a positive-sense, single-stranded enveloped coronavirus, leads to the COVID-19 disease and has been shown to be neurotropic and epitheliotropic and enter through the nasal cavity. This allows the virus to infect the neuroepithelium and bind to the ACE2-receptor rich areas of the nasal mucosa. Hyposmia and hypogeusia has been shown to affect a vast majority of people who acquire COVID-19 (45,48). The plausible mechanisms of the pathogenesis of anosmia are olfactory epithelium neuronal receptor damage directly by the virus, or cytokine storm leading to secondary damage, potentially to Bowman's gland cells and sustentacular cells after infection of supporting and perivascular cells (45). Brain imaging has shown that after SARS-CoV-2 infection, predominant changes in the brain were increased markers of tissue damage and reduction of gray matter in regions functionally relevant to the primary olfactory cortex (48) (Figure 3.5). Anosmia and ageusia can resolve after recovery, although 11% and 9% experience loss of smell and taste long-term, respectively, as post-acute sequelae of SARS-CoV-2 infection (49).

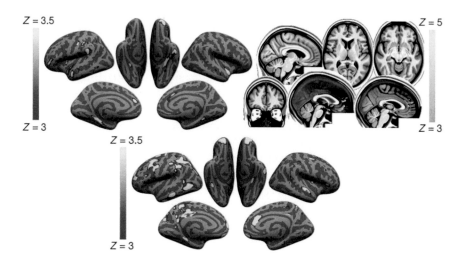

Figure 3.5 Gray matter changes between control and COVID-19 brains showing reductions in the parahippocampal gyrus and orbitofrontal cortex, which have direct connections with the olfactory bulb, as well as the anterior cingulate cortex and temporal pole, supramarginal gyrus, and insula, which is the primary gustatory cortex.

Source: Reprinted with permission from (48).

3.6 ALTERED CONSCIOUSNESS AND NEUROPSYCHIATRIC MANIFESTATIONS

Many infections lead to neuropsychiatric symptoms and impaired consciousness, such as drowsiness, fatigue, lethargy, delirium, confusion, hallucinations, depression, apathy, and coma. Several RNA viruses can cause impaired consciousness, including coronavirus, influenza, Nipah virus, the encephalitis-causing viruses, and others (4,5,14,35). These are common symptoms of toxic-metabolic encephalopathy, seizures, parenchymal damage, and neurodegeneration (12). Although the exact mechanisms are unknown, potential hypotheses include direct effect of viruses on the CNS or inflammation and subsequent microglial dysfunction (12,36,49,50).

Nipah virus, a negative-sense, single-stranded enveloped RNA virus, can cause drowsiness, disorientation, and confusion, which may be attributed to its ability to infect neurons and microglia that express CD68 in animal models (35). COVID-19 also exhibits fatigue and cognitive dysfunction, colloquially referred to as "brain fog," that can persist beyond recovery as post-acute sequelae (3,50). Apathy and chronic fatigue were found as the most common post-acute sequelae of SARS-CoV-2 infection, and impaired consciousness was reported in 37% of hospitalized patients (12,29,36). Patients who are hospitalized may be more likely to display delirium and confusion, as the risk factors are advanced age, comorbidities, severe illness, and malnutrition. Severe illness can lead to sepsis and the release of cytokines contributing to encephalopathy, such as IL-6, IL-8, IL-10, and TNF-α, which have all been implicated in states of confusion and delirium (12,36). Coronavirus has also shown microglia-mediated upregulation of proinflammatory signals (36).

Microglia can play an important role in viral-mediated neuropsychiatric conditions and fatigue. Chronic stress, like that seen in the COVID-19 pandemic, can lead to dystrophic microglia prone to phagocytosis and apoptosis, as well as greater microglial reactivity. This and microglia- and astrocyte-induced neuroinflammation are risk factors for depression, schizophrenia, and cognitive decline (36). Furthermore, COVID-19 post-acute sequelae is symptomatically similar to myalgic encephalomyelitis/chronic fatigue syndrome (ME/CFS), a condition of chronic or relapsing idiopathic fatigue, which could possibly be caused by a neuroinvasive, non-pathogenic virus (49). ME/CFS may result from stress-induced microglial activation that leads to neuroinflammation in the hypothalamus, potentially offering some insight into the pathogenesis of neurocognitive symptoms following viral infections (50).

3.7 SEIZURES, STATUS EPILEPTICUS, AND EPILEPSY

A **seizure** is a burst of sudden, uncontrolled events of electrical activity in the brain. They can be confined to one area of the brain as **focal seizures** or encompass the whole brain as **generalized seizures** and can present as convulsions or loss of consciousness or awareness. Infections of the CNS are one of the leading causes of seizures, along with trauma, stroke, tumors, and any structural or functional abnormalities in the brain (8,51). Hyperthermia can lead to excitation in the brain and **febrile seizures**, which are seizures triggered by high fever and the most common type of convulsions in children (8,11). Some of these seizures can result in febrile status epilepticus (52). **Status epilepticus** is defined as a continuous seizure lasting longer than 5 minutes or recurrent seizures without returning to baseline in between. One episode of status epilepticus can greatly increase the risk of developing **epilepsy**, a susceptibility to having recurrent unprovoked seizures (8,52). RNA viruses known to lead to febrile seizures include alphaviruses, flaviviruses, enteroviruses, parechoviruses, influenza, parainfluenza viruses, Nipah virus, chikungunya virus, Toscana virus, measles, mumps HIV, rotavirus, respiratory syncytial virus, SARS-CoV-2, and human metapneumovirus (11,51,52). About 31% of CNS infections lead to seizures, which can make up over 50% of viral-associated neurological complications (4,5,8,11,51). Some viruses, such as rotavirus and respiratory syncytial virus, can cause seizures without fever (**afebrile seizures**) (51).

Viral infections that lead to encephalitis or systemic inflammation can exhibit **acute symptomatic seizures** within the early stages of infection in half of all patients (52). Damage to the BBB in encephalitis results in a downregulation of potassium channels, aquaporins, and glutamate transporters in astrocytes, creating the condition for excess activation of NMDA glutamate receptors and neuronal hyperexcitability (51,52). Furthermore, seizures can correspond to the number of infiltrating macrophages, which can promote proinflammatory cytokines IL-6 and TNF-α that disrupt the neuronal excitation/inhibition balance and lead to neuronal and glial apoptosis (14,53). Four percent of encephalitis patients develop status epilepticus, and **late unprovoked seizures** can also develop at any point after recovery (52).

Acute symptomatic seizures generally do not recur but can increase the risk for late unprovoked seizures and developing epilepsy from 10% to 22%, due to potential injury and maladaptive changes in neural circuitry (11,52,54). Early seizures and inflammation can alter synaptic spine structure, balance between inhibitory and excitatory neurons and synaptic channels, and transcription involved in controlling seizure threshold, leading to encephalopathy and neuronal hyperexcitability that can persist (52). **Temporal lobe epilepsy** is the most common type of epilepsy in adults due to viral infections, particularly as a result of neuronal loss and gliosis in the hippocampus that can promote seizures (9,52) (Figure 3.6). Infarctions, hypoxic-ischemic injury, and gliosis may all lead to epileptogenic foci in the brain (8). For instance, Japanese encephalitis virus and West Nile virus generally affect subcortical structures and spare cerebral cortex, as opposed to

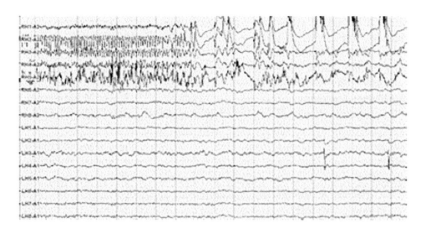

Figure 3.6 Right hippocampal seizure onset in patient with temporal lobe epilepsy.

Source: Reprinted with permission from (55).

Table 3.1 Neuropathologies and Their Associated RNA Viruses

Neuropathology	Associated RNA Viruses
Encephalopathy/ encephalitis	Influenza, rotavirus, human respiratory syncytial virus, coronavirus, human metapneumovirus, rabies virus, measles, mumps, Japanese encephalitis virus, tick-borne encephalitis virus, West Nile virus, SARS-CoV-2, Zika virus, dengue virus, St. Louis encephalitis virus, Eastern and Western equine encephalitis viruses, poliovirus
Meningitis	Mumps, measles, flaviviruses, enteroviruses
Myelitis	Enteroviruses, flaviviruses, influenza, HIV, SARS-CoV-2
Neurodegeneration	Coronaviruses, influenza, hepatitis C virus, HIV, Japanese encephalitis virus, West Nile virus, Zika virus, dengue virus, chikungunya virus
Anosmia/ageusia	Coronaviruses, influenza, parainfluenza viruses, rhinoviruses
Altered consciousness	Coronavirus, influenza, Nipah virus, the encephalitis-causing viruses
Seizures/epilepsy	Alphaviruses, flaviviruses, enteroviruses, parechoviruses, influenza, parainfluenza viruses, Nipah virus, chikungunya virus, Toscana virus, measles, mumps HIV, rotavirus, respiratory syncytial virus, SARS-CoV-2, human metapneumovirus
Stroke	SARS-CoV-2, HIV, influenza, parainfluenza, hepatitis C virus
Microcephaly	Zika virus, Japanese encephalitis virus, West Nile virus, dengue virus, influenza, HIV, poliovirus, measles
Neuropsychiatric conditions	Influenza

the temporal lobe (53). Such flaviviruses are known to cause early and late seizures (52) (Table 3.1).

3.8 STROKE AND VASCULOPATHY

Vasculopathy describes any condition affecting the blood vessels. Vasculopathies that lead to vessel wall inflammation and shape change, rupture of atherosclerotic plaques, or destabilization of cardiovascular conditions can lead to **stroke** (56,57). **Ischemic stroke** results when blood supply to the brain is blocked or stalled (Figure 3.7), and **hemorrhagic stroke** occurs when blood vessel ruptures lead to bleeding inside the brain (Figure 3.8). Strokes can be caused by atherosclerosis, hypertension, and atrial fibrillation (AF), and increasing evidence shows that viruses can increase the risk of stroke by systemic inflammation, thrombosis, or vasculitis, and directly contribute to or aggravate strokes (12,57–59). Viruses associated with stroke include SARS-CoV-2, HIV, influenza, parainfluenza, and hepatitis C virus.

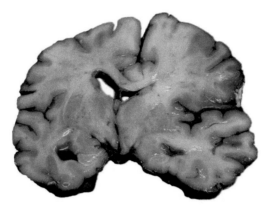

Figure 3.7 Acute ischemic stroke due to middle cerebral artery occlusion.

Source: Reprinted with permission from (60).

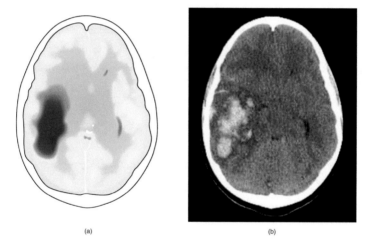

(a)   (b)

Figure 3.8 Hemorrhagic stroke, with (a) showing an artistic representation of an MRI in (b).

Source: Reprinted with permission from (61).

3.8.1 Ischemic Stroke

The risk of ischemic stroke during SARS-CoV-2 infection is about 5%, the second most common neurological manifestation of COVID-19, while hemorrhagic strokes are less common. The three main mechanisms of ischemic stroke are a hypercoagulable state, vasculitis, and cardiomyopathy (52,58).

3.8.1.1 Hypercoagulable State

Viruses can disrupt homeostasis processes that govern the balance between pro- and anticoagulation mechanisms, either directly or indirectly (56). One method is viral invasion of the vascular epithelium. Inflammation and damage to the endothelial cells can lead to further inflammation and cytokine storm, which alters the balance of local clotting factors, triggering the coagulation cascade and the formation of thromboemboli (56,58). Influenza and parainfluenza viruses can infect and activate endothelial cells, which alters the composition of phospholipids and increases inflammatory binding sites that produce procoagulant cytokines and generate thrombin (56). SARS-CoV-2 also has the ability to enter and facilitate local inflammation and endotheliitis, leading to impaired microcirculatory function and microvascular thrombosis (47,58). Furthermore, viruses can bind directly to platelets, activating them and contributing to thrombus initiation and growth (56). This hypercoagulability can lead to large vessel occlusion of small vessel stroke (56,58). In SARS-CoV-2 infection, 20%–55% of patients hospitalized show laboratory evidence of a hypercoagulable state, including increased D-dimer levels, C-reactive protein levels, ferritin levels, and lactate dehydrogenase levels, as well as elevated leukocyte count, prolongation of prothrombin time, mild thrombocytopenia and decreased fibrinogen levels consistent with disseminated intravascular coagulation (12,47,58). Further reports showed microvascular platelet fibrin-rich thrombotic depositions with the virus identified in these microischemic regions. Microvascular thrombosis and injury to the neurovascular unit can result in BBB disruption and viral translocation that leads to further CNS involvement (47,58).

3.8.1.2 Vasculitis

Vasculopathy can be inflammatory, hemostatic, metabolic, or genetic. Inflammation of vessel walls is described as **vasculitis** and can be direct or indirect (56,58). In direct vasculitis, viruses infecting vessels directly cause structural damage to the vascular wall. Binding to endothelial cells can activate adhesion molecules, releasing proinflammatory cytokines and chemoattractants (56). Endothelial dysfunction and cytokine storm can lead to changes in the shape of blood vessel walls characterized by vasculitis, and immune complexes synthesized in blood vessel walls can also cause vasculitis (56,58). Furthermore, rupture of atherosclerotic plaques can result from an acceleration of atherosclerosis development or destabilization of existing plaques. HIV, for instance, is associated with vasculitis, and some antiretroviral therapies are believed to accelerate atherosclerosis (56,57,62). Infection can lead to atheroma development due to inflammation and direct enzymatic degradation of the atheromatous plaque cap. Systemic inflammatory reaction degrades and thins the plaque and results in expansion in cholesterol from liquid to solid, which can create sharp crystals that perforate the fibrous cap. This can result in eventual plaque rupture, which leads to turbulence in cerebral blood flow and a highly thrombogenic site (56). HIV-macrophages in mice were shown to have impaired cholesterol efflux, which is associated with being highly atherogenic (63). An autoimmune mechanism can also be at play, where shared epitopes between the pathogen and the host can lead to destruction of the vessel wall (56). An inflammatory response to viruses like HIV can lead to a targeting of self-antigens (57).

Indirect vasculitis is associated with changes in the renin-angiotensin system (47,56). These changes can be due to alterations in ACE and ACE2. Vessels may be directly inflamed by direct effect on ACE2, the receptors for SARS-CoV-2 viral entry, and lead to increased risk for ischemic stroke, so viral antibodies may be formed against ACE2, causing it to be downregulated (56,58). ACE overactivation and ACE2 underactivation therefore leads to vasoconstriction, arteriopathy, and a prothrombotic state that can cause cerebral ischemia (56). In SARS-CoV-2, this effect may show synergy with other mechanisms, as well as hypoxia and BBB damage, uniquely making younger people more at risk for ischemic stroke (47,58). Additionally, HIV is the second leading risk factor for hypertension and the most important one among young stroke patients (57). HIV and antiretroviral therapy may result in deficient generation of endothelium-dependent nitric oxide. Impaired compensatory responses by the vasculature to reduce blood pressure can lead to reduction in cerebral blood flow and increase the risk of stroke (62). HIV-infected patients may also be immunosuppressed and more susceptible to opportunistic infections that can drive further vessel wall inflammation. Interestingly, large artery disease occurs more frequently in HIV patients, as does basal ganglia ischemia (57).

3.8.1.3 Cardiomyopathy

Cardiomyopathy, a disease of the heart muscle, and myocardial dysfunction can lead to thromboembolisms that cause ischemic stroke (47). Infection, inflammation, and hypercoagulability can all lead to changes in cardiac function. For instance, viruses can be associated with myocarditis and increase the risk for a cardioembolic event (56). Direct invasion of cardiomyocytes by viruses can cause myocarditis, leading to injury and death of the cells. In COVID-19, this may be related to viral entry through ACE2 receptors and the dysfunction caused by its downregulation (58). ACE2 knockout in mouse models leads to severe left ventricular (LV) dysfunction, which is an important cause of LV thrombi and cardioembolism that significantly increases the risk of stroke (47). Myocarditis may also be caused by the inflammatory response and cytokine storm caused by viral infection, which can lead to microvascular thrombosis and myocardial dysfunction

(47,58). Inflammatory cytokines may further affect cardiomyocyte function related to tachycardia and arrhythmias (47). Stimulation of the sympathetic nervous system and increased cardiac output leads to stress-induced AF that increases risk of cardioembolic stroke or changes in cerebral perfusion that exceed autoregulatory capacities (47,56,58). Cardiomyopathy can also be caused by respiratory failure and hypoxemia from infection, or prolonged infection can lead to sepsis (56,58). Sepsis can lead to dehydration, which results in blood pressure fluctuations and changes in cerebral blood perfusion pressure. The lack of normal blood flow, or **stasis**, can lead to prothrombotic states that increase the risk for thromboembolism, AF, and stroke (47,56). SARS-CoV-1 exhibited cardioembolic mechanisms in 60% of cases (58).

3.8.2 Hemorrhagic Stroke

Viral infection of vascular endothelial cells and damage to the vasculature can also increase the risk of hemorrhagic infarcts (12). Some of these may occur secondary to ischemic stroke. For instance, influenza virus can disrupt the BBB and increase the rate and size of intracerebral hemorrhages after ischemic stroke is reperfused (59). This may be due to the release of cytokines aggravating ischemic brain damage, causing hemorrhagic encephalopathy (58,59). The inflammatory cascade may lead to BBB breakdown due to cerebral edema, raised intracranial pressure, cerebral herniation, and subsequent infarction (11). In COVID-19, a direct mechanism is possible, as the affinity of the virus for ACE2 receptors can damage intracranial arteries by infecting endothelial and arterial smooth muscle cells, eventually leading to vessel wall rupture (58).

3.8.3 Headache

Although mechanisms are unclear, headache is a common symptom of several viruses and may be related to the same functions that lead to stroke, such as direct viral injury, inflammatory process, hypoxemia, coagulopathy, and endothelial involvement (64). However, it is also possible that persistent activation of the immune system and trigeminovascular activation play a role in the onset of headache (65).

3.9 FETAL NEURODEVELOPMENT COMPLICATIONS

Viral infections during pregnancy can have adverse effects on fetuses in the important stages of development. Neurodevelopment begins prenatally in the third week of gestation and continues postnatally until adulthood, and disruptions early in this process can lead to disorders, including but not limited to microcephaly, schizophrenia, autism spectrum disorder, and epilepsy

(66,67). Several RNA viruses have been linked to neurodevelopmental complications, including Zika virus, Japanese encephalitis virus, West Nile virus, dengue virus, influenza, HIV, poliovirus, and measles (52,66).

3.9.1 Microcephaly and Brain Anomalies

Microcephaly is a condition of abnormally small head circumference typically associated with incomplete brain development (68) (Figure 3.9). While it can be caused by genetic defects or environmental perturbations, pathogens such as RNA viruses have been shown to influence such growth restrictions and brain abnormalities, either directly or indirectly through inflammatory response (66). Other brain abnormalities that can result from RNA viruses include **lissencephaly** (smooth brain, without folds), **ventriculomegaly** (large ventricles), **hydrocephalus** (CSF buildup in ventricles), brain calcifications, cerebellar hypoplasia, cortical malformations, abnormal vasculature, axonal

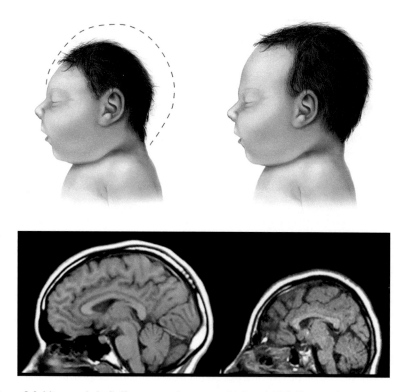

Figure 3.9 Microcephaly (*left*) compared to normal infant (*right*). *Top panel*: an artistic representation comparing the physical features of microcephaly to a normal infant. *Bottom panel*: MRI of a case of microcephaly in comparison to normal infant MRI.

Source: Reprinted with permission from (69).

rarefaction, and an increase and change in the morphology and behavior of astrocytes (**astrogliosis**) (68,70,71).

Zika virus (ZIKV) is the first flavivirus associated with congenital defects, being able to pass from placenta to fetus (52). In this way, ZIKV can directly pass through the fetal BBB, which is more permeable during development, through transcytotic or paracellular pathways without damaging the integrity (66,67). Once in the CNS, ZIKV can infect astrocytes, microglia, and neural progenitor cells (NPCs), which are multipotent stem cells that differentiate into neurons, astrocytes, and oligodendrocytes (67). It has been established that ZIKV is toxic and disruptive to NPCs (67,71). Not only can the virus downregulate genes regulating cell cycle, cell division, neurogenesis, and axonal guidance and differentiation, but it can also induce cell death through ER stress, unfolded protein response, apoptosis, pyroptosis, and autophagy (67,71,72). This cell cycle arrest and uncontrolled apoptosis of NPCs can lead to the decreased brain volume and disorganization of cortical neurons consistent with microcephaly (67).

Furthermore, ZIKV can also elicit a strong immune response that can disrupt fetal neurodevelopment. While symptoms can be mild in adults, the proinflammatory cytokines released can reach the fetal brain and alter NPC biology, activate microglia, induce astrogliosis, and promote further neuroinflammation (67). ZIKV can dysregulate genes involved in the immune response, with IL-1β and TNF-α expression dramatically increasing (71). This immune response can also lead to abnormal development of the neurovasculature. ZIKV infection can lead to abnormal vascular density and increased vessel diameter in the brain, in addition to downregulation of tight junction proteins, which results in a leaky BBB (66,71,72). BBB leakage can lead to brain calcifications, particularly in the basal ganglia and junction between cortical and subcortical white matter, as well as extensive brain damage and neuronal death (70,71). All of these can lead to postnatal characteristics similar to microcephaly and other malformations.

3.9.2 Schizophrenia and Neuropsychiatric Disorders

Viral infection during pregnancy has also been linked to **schizophrenia** in offspring, a neuropsychiatric disorder that affects a person's interpretation of reality. One potential mechanism would be the resultant localized nonlethal apoptosis of progenitors, which causes neurite and synaptic loss that change neural circuit formation and function (66). Another mechanism is related to the maternal inflammatory response, which has been characterized in influenza infection, where the virus does not appear in the placenta or fetal brain (66,67). Proinflammatory cytokines can cause fetal neuroinflammation and increase risk of schizophrenia (67). When proinflammatory cytokines such as IL-1β and TNF-α are introduced to the brain, crossing the BBB, altered cognition is noted. In combination with IFN-γ, these can induce the expression of MHC1, which has the potential to downregulate synapses when activated. TNF-α, IL-8, and C-reactive protein in maternal serum are also linked to schizophrenia risk in offspring (66). This is further validated by elevated anti-inflammatory cytokines in diagnosed individuals, such as IL-6, which can alter cognitive behavior in progeny. Several pro- and anti-inflammatory cytokines, like IL-6, TNF-α, IL-1β, and IFN-β, have been shown to cause fetal growth restriction by hypoxia (66,67). The result points to offspring with reduced brain volumes, disturbed neuronal migration and expression, neurite and synapse loss, and no astrogliosis, which can be indicative of behavior and cognition impairments in adulthood (67). Similar mechanisms have been described for bipolar disorder and major depressive disorder. Interestingly, studies suggest that male fetuses are more susceptible to these conditions, due to less resistance to intrauterine stress and a faster microglial maturation pathway (66).

3.9.3 Autism Spectrum Disorder

Similarly, changes in brain size, neural circuit formation, and cortical structure may also result in **autism spectrum disorder** (ASD), a neurodevelopmental disorder that can affect cognition, behavior, and social interactions. In considering viral effects on brain size, viral infection can reduce volumes of several brain regions that can lead to impairments in exploratory behavior and social interaction (67). However, viral effects on programmed cell death can also change circuit connectivity and result in an abnormal expansion of neocortical excitatory neurons over inhibitory neurons, which is reflected in ASD with an enlarged brain phenotype (**macrocephaly**) in early postnatal years. This enhanced excitatory synaptic connectivity has been similarly observed in mice defective for C1q and CX3CR1, which usually moderate synaptic pruning (66). Viruses can likewise sequester or inactivate the complement cascade as an evasion strategy. Activated microglia can secrete complement components that result in abnormal synaptic pruning and can expose the fetal brain to neuronal loss due to proinflammatory state. Proinflammatory cytokines in the brain like IL-1β and TNF-α have been associated with social withdrawal, and elevated anti-inflammatory cytokines in serum are associated with alterations in cognitive behavior and working memory (66,67). For example, high levels of IL-6 have been linked to an altered front limbic circuit, as well as greater amygdala size and connectivity, affecting regions of the brain involved in sensory processing and integration, salience deletion, learning and memory, and cognitive ability (67). Because male offspring have a faster microglial maturation pathway, affecting expression of immune-related genes, they are also more susceptible to ASD and related conditions, such as attention-deficit/hyperactivity disorder (66).

3.9.4 Epilepsy

While epilepsy can arise in childhood or adulthood, viral infections can also generate epilepsy in utero. For instance, epilepsy is one of the main neurological outcomes of congenital Zika syndrome, reported in almost 60% of infants in the first 4 months of life (52,68). ZIKV infection can release cytotoxic factors like IL-1β, TNF-α, and glutamate, which can influence epileptic seizures (71). Dysfunctional BBB as a result of viral infection can also be implicated in seizures and epilepsy (66). Onset of epilepsy and other involuntary movement disorders can be correlated to the degree of brain damage from viruses like ZIKV (70).

REFERENCES

1. (NINDS) NI of ND and S. Definitions 2018.
2. Virhammar J, Kumlien E, Fällmar D, Frithiof R, Jackmann S, Sköld MK, et al. Acute necrotizing encephalopathy with SARS-CoV-2 RNA confirmed in cerebrospinal fluid. *Neurology* 2020;95(10); 445–449.
3. Sanchez CV, Theel E, Binnicker M, Toledano M, McKeon A. Autoimmune encephalitis after SARS-CoV-2 infection. *Neurology* 2021;97(23); e2262–e2268.
4. Bohmwald K, Gálvez NMS, Ríos M, Kalergis AM. Neurologic alterations due to respiratory virus infections. *Frontiers in Cellular Neuroscience* 2018;12; 386.
5. Liang CY, Yang CH, Lin JN. Focal encephalitis, meningitis, and acute respiratory distress syndrome associated with influenza a infection. *Medical Principles and Practice* 2018;27(2); 193–196.
6. Cappel R, Thiry L, Clinet G. Viral antibodies in the CSF after acute CNS infections. *Formerly Archives of Neurology* 1975;32(9); 629–631.
7. Deigendesch N, Stenzel W. Chapter 17 Acute and chronic viral infections. *Handbook of Clinical Neurology* 2017;145; 227–243.
8. Vezzani A, Fujinami RS, White HS, Preux PM, Blümcke I, Sander JW, et al. Infections, inflammation and epilepsy. *Acta Neuropathologica* 2016;131(2); 211–234.
9. Hellerhoff. *Mild Encephalitis with Reversible Lesion in the Splenium 6jm Noro-MRT T2ax und Sag* [Internet] 2016. Available from: https://upload.wikimedia.org/wikipedia/commons/1/13/Mild_encephalitis_with_reversible_lesion_in_the_splenium_6jm_Noro-MRT_T2ax_und_sag_-_001.jpg
10. Dubey D, Pittock SJ, Kelly CR, McKeon A, Lopez-Chiriboga AS, Lennon VA, et al. Autoimmune encephalitis epidemiology and a comparison to infectious encephalitis: Autoimmune encephalitis. *Annals of Neurology* 2018;83(1); 166–177.
11. SINGHI P. Infectious causes of seizures and epilepsy in the developing world. *Developmental Medicine and Child Neurology* 2011;53(7); 600–609.
12. Zubair AS, McAlpine LS, Gardin T, Farhadian S, Kuruvilla DE, Spudich S. Neuropathogenesis and neurologic manifestations of the coronaviruses in the age of coronavirus disease 2019. *JAMA Neurology* 2020;77(8); 1018–1027.
13. Morichi S, Kawashima H, Ioi H, Yamanaka G, Kashiwagi Y, Hoshika A, et al. Classification of acute encephalopathy in respiratory syncytial virus infection. *Journal of Infection and Chemotherapy* 2011;17(6); 776–781.
14. Mizuguchi M, Yamanouchi H, Ichiyama T, Shiomi M. Acute encephalopathy associated with influenza and other viral infections. *Acta Neurologica Scandinavica* 2007;115(s186); 45–56.
15. Glaser CA, Winter K, DuBray K, Harriman K, Uyeki TM, Sejvar J, et al. A population-based study of neurologic manifestations of severe influenza A(H1N1)pdm09 in California. *Clinical the Infectious Diseases Society of America*. 2012;55(4); 514–520.
16. Kawashima H, Kashiwagi Y, Ioi H, Morichi S, Oana S, Yamanaka G, et al. Production of chemokines in respiratory syncytial virus infection with central nervous system manifestations. *Journal of Infection and Chemotherapy* 2012;18(6); 827–831.

17. Tavčar P, Potokar M, Kolenc M, Korva M, Avšič-Županc T, Zorec R, et al. Neurotropic viruses, astrocytes, and COVID-19. *Frontiers in Cellular Neuroscience* 2021;15; 662578.

18. Yadav P, Chakraborty P, Jha NK, Dewanjee S, Jha AK, Panda SP, et al. Molecular mechanism and role of Japanese encephalitis virus infection in central nervous system-mediated diseases. *Viruses* 2022;14(12); 2686.

19. Scott TP, Nel LH. Lyssaviruses and the fatal encephalitic disease rabies. *Frontiers in Immunology* 2021;12; 786953.

20. Logan SAE, MacMahon E. Viral meningitis. *BMJ* 2008;336(7634); 36.

21. Gupta RK, Best J, MacMahon E. Mumps and the UK epidemic 2005. *BMJ* 2005;330(7500); 1132.

22. Mihai C, Jubelt B. Infectious myelitis. *Current Neurology and Neuroscience Reports* 2012;12(6); 633–641.

23. Mehndiratta MM, Mehndiratta P, Pande R. Poliomyelitis. *Neurohospitalist* 2014;4(4); 223–239.

24. Chauhan L, Matthews E, Piquet AL, Henao-Martinez A, Franco-Paredes C, Tyler KL, et al. Nervous system manifestations of arboviral infections. *Current Tropical Medicine Reports* 2022;9(4); 107–118.

25. Ford D, Ropka S, Collins G, Jubelt B. The neuropathology observed in wild-type mice inoculated with human poliovirus mirrors human paralytic poliomyelitis. *Microbial Pathogenesis*. 2002;33(3); 97–107.

26. Ali L, Mohammed I, Zada Y, Salem H, Iqrar A. COVID-19-associated acute transverse myelitis: A case series of a rare neurologic condition. *Cureus* 2021;13(10); e18551.

27. Zhou L, Miranda-Saksena M, Saksena NK. Viruses and neurodegeneration. *Journal of Virology* 2013;10(1); 172.

28. Chen WW, Zhang X, Huang WJ. Role of neuroinflammation in neurodegenerative diseases (Review). *Molecular Medicine Reports* 2016;13(4); 3391–3396.

29. Müller-Schiffmann A, Trossbach SV, Lingappa VR, Korth C. Viruses as "truffle hounds": Molecular tools for untangling brain cellular pathology. *Trends in Neurosciences* 2020;44(5); 352–365.

30. Lassmann H. Multiple sclerosis pathology. *Cold Spring Harbor Perspectives in Medicine* 2018;8(3); a028936.

31. Burks JS, DeVald BL, Jankovsky LD, Gerdes JC. Two coronaviruses isolated from central nervous system tissue of two multiple sclerosis patients. *Science* 1980;209(4459); 933–934.

32. Kristensson K, Norrby E. Persistence of RNA viruses in the central nervous system. *Annual Review of Microbiology* 1986;40(1); 159–184.

33. Yeh EA, Collins A, Cohen ME, Duffner PK, Faden H. Detection of coronavirus in the central nervous system of a child with acute disseminated encephalomyelitis. *Annual Review of Microbiology* 2004;113(1); e73–e76.

34. Talbot PJ, Arnold D, Antel JP. The Mechanisms of neuronal damage in virus infections of the nervous system. *Current Topics in Microbiology* 2012;253; 247–271.

35. Chakravarty N, Senthilnathan T, Paiola S, Gyani P, Cario SC, Urena E, et al. Neurological pathophysiology of SARS-CoV-2 and pandemic potential RNA viruses: A comparative analysis. *FEBS Letters* 2021;595(23); 2854–2871.

36. Awogbindin IO, Ben-Azu B, Olusola BA, Akinluyi ET, Adeniyi PA, Paolo TD, et al. Microglial implications in SARS-CoV-2 infection and COVID-19: Lessons from viral RNA neurotropism and possible relevance to Parkinson's disease. *Frontiers in Cellular Neuroscience* 2021;15; 670298.

37. Cohen ME, Eichel R, Steiner-Birmanns B, Janah A, Ioshpa M, Bar-Shalom R, et al. A case of probable Parkinson's disease after SARS-CoV-2 infection. *Lancet Neurology* 2020;19(10); 804–805.

38. Marogianni C, Sokratous M, Dardiotis E, Hadjigeorgiou GM, Bogdanos D, Xiromerisiou G. Neurodegeneration and inflammation—An interesting interplay in Parkinson's disease. *International Journal of Molecular Sciences* 2020;21(22); 8421.

39. Mandel SA, Morelli M, Halperin I, Korczyn AD. Biomarkers for prediction and targeted prevention of Alzheimer's and Parkinson's diseases: Evaluation of drug clinical efficacy. *EPMA Journal* 2010;1(2); 273–292.

40. Sivadon-Tardy V, Orlikowski D, Porcher R, Sharshar T, Durand M, Enouf V, et al. Guillain-Barré Syndrome and Influenza virus infection. *Clinical Infectious Diseases* 2009;48(1); 48–56.

41. Trujillo JR, Jaramillo-Rangel G, Ortega-Martinez M, Oliveira Acp de, Vidal JE, Bryant J, et al. International NeuroAIDS: Prospects of HIV-1 associated neurological complications. *Cell Research* 2005;15(11–12); 962–969.

42. Shapshak P, Kangueane P, Fujimura RK, Commins D, Chiappelli F, Singer E, et al. Editorial Neu-roAIDS review. *Aids* 2011;25(2); 123–141.
43. Miller RG, Parry GJ, Pfaeffl W, Lang W, Lippert R, Kiprov D. The spectrum of peripheral neuropathy associated with ARC and AIDS. *Muscle & Nerve* 1988;11(8); 857–863.
44. Rezaie A, Parmar R, Rendon C, Zell SC. HIV-associated vacuolar myelopathy: A rare initial presenta-tion of HIV. *Sage Open Medical Case Reports* 2020;8; 2050313X20945562
45. Othman BA, Maulud SQ, Jalal PJ, Abdulkareem SM, Ahmed JQ, Dhawan M, et al. Olfactory dysfunc-tion as a post-infectious symptom of SARS-CoV-2 infection. *Annals of Medicine and Surgery* 2022;75; 103352.
46. Altable M, Serna JM de la. Neuropathogenesis in COVID-19. *Journal of Neuropathology & Experi-mental Neurology* 2020;79(11); 1247–1249.
47. Sagris D, Papanikolaou A, Kvernland A, Korompoki E, Frontera JA, Troxel AB, et al. COVID-19 and ischemic stroke. *European Journal of Neurology* 2021;28(11); 3826–3836.
48. Douaud G, Lee S, Alfaro-Almagro F, Arthofer C, Wang C, McCarthy P, et al. SARS-CoV-2 is associated with changes in brain structure in UK Biobank. *Nature* 2022; 1–17.
49. Moghimi N, Napoli MD, Biller J, Siegler JE, Shekhar R, McCullough LD, et al. The neurological manifestations of post-acute sequelae of SARS-CoV-2 infection. *Current Neurology and Neuroscience Reports* 2021;21(9); 44.
50. Theoharides TC, Cholevas C, Polyzoidis K, Politis A. Long-COVID syndrome-associated brain fog and chemofog: Luteolin to the rescue. *Biofactors* 2021;47(2); 232–241.
51. Libbey JE, Fujinami RS. Neurotropic viral infections leading to epilepsy: Focus on theilers murine encephalomyelitis virus. *Future Virology* 2011;6(11); 1339–1350.
52. Löscher W, Howe CL. Molecular mechanisms in the genesis of seizures and epilepsy associated with viral infection. *Frontiers in Molecular Neuroscience* 2022;15; 870868.
53. Zhang P, Yang Y, Zou J, Yang X, Liu Q, Chen Y. Seizures and epilepsy secondary to viral infection in the central nervous system. *Frontiers in Molecular Neuroscience* 2020;2(1); 12.
54. Annegers JF, Hauser WA, Beghi E, Nicolosi A, Kurland LT. The risk of unprovoked seizures after encephalitis and meningitis. *Neurology* 1988;38(9); 1407–1410.
55. F V, D T, M M de O, AL V. Neuromodulation of bilateral hippocampal foci, an alternative for mesial temporal lobe seizures in patients with non-lesional MRI: Long-term follow-up. *Epilepsy Journal* 2016;2(4); 1–5.
56. Bahouth MN, Venkatesan A. Acute viral illnesses and ischemic stroke. *Stroke* 2021;52(5); 1885–1894.
57. Benjamin LA, Allain TJ, Mzinganjira H, Connor MD, Smith C, Lucas S, et al. The role of human immunodeficiency virus–associated vasculopathy in the etiology of stroke. *Journal of Infectious Dis-eases* 2017;216(5); 545–553.
58. Spence JD, Freitas GR de, Pettigrew LC, Ay H, Liebeskind DS, Kase CS, et al. Mechanisms of stroke in COVID-19. *Cerebrovascular Diseases* 2020;49(4); 451–458.
59. Muhammad S, Haasbach E, Kotchourko M, Strigli A, Krenz A, Ridder DA, et al. Influenza virus infec-tion aggravates stroke outcome. *Stroke* 2011;42(3); 783–791.
60. 101 M. MCA-Stroke-Brain-Human-2 [Internet] 2006. Available from: https://upload.wikimedia.org/wikipedia/commons/3/36/MCA-Stroke-Brain-Human-2.JPG
61. College O. *1602 The Hemorrhagic Stroke-02* [Internet] 2013. Available from: https://commons.wiki-media.org/wiki/File:1602_The_Hemorrhagic_Stroke-02.jpg
62. Sen S, Rabinstein AA, Elkind MSV, Powers WJ. Recent developments regarding human immunodefi-ciency virus infection and stroke. *Cerebrovascular Diseases* 2012;33(3); 209–218.
63. Nagel MA, Mahalingam R, Cohrs RJ, Gilden D. Virus vasculopathy and stroke: An under-recognized cause and treatment target. *Infectious Disorders – Drug Targets* 2010, Apr;10(2); 105–111.
64. Rocha-Filho PAS. Headache associated with COVID-19: Epidemiology, characteristics, pathophysiol-ogy, and management. *Headache: The Journal of Head and Face Pain* 2022;62(6); 650–656.
65. Tana C, Bentivegna E, Cho SJ, Harriott AM, García-Azorín D, Labastida-Ramirez A, et al. Long COVID headache. *Journal of Headache and Pain* 2022;23(1); 93.
66. Ganguli S, Chavali PL. Intrauterine viral infections: Impact of inflammation on fetal neurodevelop-ment. *Front Neurosci-switz* 2021;15; 771557.
67. Elgueta D, Murgas P, Riquelme E, Yang G, Cancino GI. Consequences of viral infection and cytokine production during pregnancy on brain development in offspring. *Frontiers in Immunology* 2022;13; 816619.

68. Oliveira-Filho J, Felzemburgh R, Costa F, Nery N, Mattos A, Henriques DF, et al. Seizures as a complication of congenital Zika syndrome in early infancy. *American Journal of Tropical Medicine and Hygiene* 2018;98(6); 1860–1862.

69. Evolutionary history of a gene controlling brain size. *PLoS Biology* 2004;2(5); e134. Available from: https://journals.plos.org/plosbiology/article?id=10.1371/journal.pbio.0020134

70. Pessoa A, Linden V van der, Yeargin-Allsopp M, Carvalho MDCG, Ribeiro EM, Braun KVN, et al. Motor abnormalities and epilepsy in infants and children with evidence of congenital Zika virus infection. *Pediatrics* 2018;141(Suppl 2); S167–S179.

71. Shao Q, Herrlinger S, Yang SL, Lai F, Moore JM, Brindley MA, et al. Zika virus infection disrupts neurovascular development and results in postnatal microcephaly with brain damage. *Development* 2016;143(22); 4127–4136.

72. Christian KM, Song H, Ming G li. Pathophysiology and mechanisms of Zika virus infection in the nervous system. *Annual Review of Neuroscience* 2019;42(1); 249–269.

Part II

MOLECULAR AND CLINICAL OUTCOMES ASSOCIATED WITH RNA VIRUS INFECTIONS IN THE HUMAN BRAIN

Chapter 4

Molecular and Cellular Alterations Underlying Neurological Outcomes in HIV Infection

Sabyasachi Dash

4.1 INTRODUCTION

As described in the previous chapters, HIV virus is a retrovirus that contains two copies of positive ssRNA genome (Figure 4.1).

HIV infects immune cells such as CD4+ T cells, macrophages, and microglial cells. HIV-1 virus entry to macrophages and CD4+ T cells is mediated through interaction of the virion envelope glycoproteins (gp120) with the CD4 molecule expressed on the target cells' membrane and with chemokine co-receptors such as CXCR4, and CCR5. Risky behaviors including unprotected sexual intercourse and the exchange of syringes and body fluids are some of the major modes of HIV transmission among individuals. Studies with both same-sex and opposite-sex couples have shown that HIV cannot be transmitted through unprotected sexual intercourse if the HIV-positive partner has a undetectable viral load (1,2). On the other hand, non-sexual transmission can occur from an infected mother to her infant during pregnancy, during childbirth by exposure to her blood or vaginal fluid, or through breast milk (3,4). This is because the HIV virus is present as both free virus particles and virus within infected immune cells in these bodily fluids. In summary, HIV attacks the immune system by destroying specific white blood cells called CD4 positive (CD4+) T cells that are vital to the immune system. This causes a shortage of these immune cells, thus a weakened immune system that leaves HIV-positive patients vulnerable to other infections (also called co-infections) and prone to secondary

DOI: 10.1201/9781003285823-4

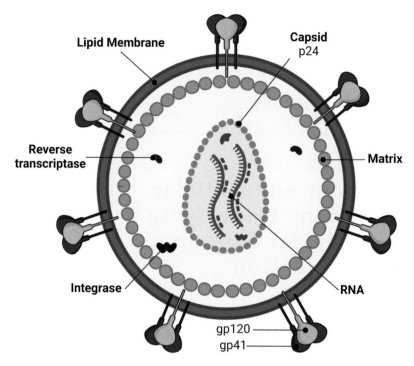

Figure 4.1 General molecular architecture of the human immunodeficiency virus (HIV).

diseases and additional complications. HIV infection is aggressive and always causes progressive immune system destruction, that can be seen by dramatic reduction in the numbers of CD4+ and CD8+ T cells, which occurs much more rapidly when patients do not receive any treatment (5). Ultimately, this leads to acquired immune deficiency syndrome (AIDS), which is characterized by susceptibility to various opportunistic infections. AIDS is diagnosed when the immune system composition is altered below a threshold percentage or when T cell quantity is too low. AIDS is the final stage of HIV infection featuring a high rate of opportunistic/co-infections, such as pneumonia or tuberculosis, that do not typically affect people with healthy immune systems. Although HIV infection and AIDS primarily affect the immune system, they also disturb the nervous system and can lead to a wide range of severe neurological disorders, particularly if HIV goes untreated and progresses to AIDS. Even though HIV does not directly invade nerve cells (neurons), research has established that the virus impacts their function by secondary mechanisms such as by infecting cells called glia, resident immune cells in the brain that express the necessary receptors and co-receptors and play an integral role in the structure and function of the brain.

4.2 HIV-ASSOCIATED CENTRAL NERVOUS SYSTEM INJURY

The central nervous system (CNS) has been established as a major target of HIV infection (6–8). Strikingly, the loss of neuronal population in HIV-infected patients is a secondary event to infection of microglia and resident immune cells in the brain. It is postulated that the free virus particles may penetrate the CNS by crossing the capillary endothelial cells that are integral to the blood–brain barrier (BBB) or can be trafficked into brain by infected lymphocytes or monocytes (Figure 4.2). Once a CNS infection is established,

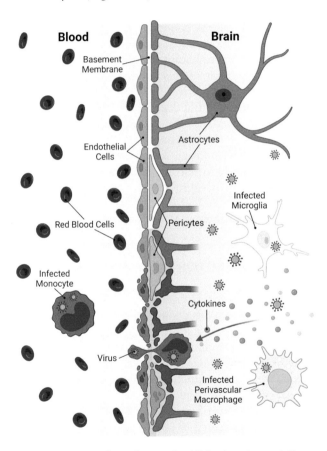

Figure 4.2 The classic "Trojan horse" mechanism for HIV-induced neuroinflammation via blood–brain barrier (BBB) disruption. Receptors on CD4$^+$ T cells or monocytes (*red*) can bind to HIV, allowing for its replication and circulation in the host system. The BBB expresses chemokines with chemokine-specific receptors for recruitment of immune cells. As a result, HIV can efficiently cross the otherwise impermeable BBB utilizing these infected circulating T cells and monocytes and synergistically impairing the structure of tight junction proteins that tightly bind the microvascular endothelial cells leading to a successful invasion and establishment of productive infection in the CNS.

neuronal injury likely occurs via indirect mechanisms such as toxicity from virus proteins, macrophage factors, cytokines, and chemokines, or due to a loss of neurotrophic factors (9–16).

HIV infection results in disruption of neuronal function by a variety of mechanisms that can be grouped into three broad categories: viral factors, host factors, and co-factors. Viral factors derive from the virus itself and include several proteins encoded by the viral genome. On the other hand, host factors evolve indirectly from infection but may damage even uninfected cells such as neurons. There is also the influence of co-factors or comorbid conditions that may contribute to enhance the pathogenicity of HIV. In the nervous system, these factors converge in producing damage to the elaborate network of connections between neurons that take place at dendrites and synapses. This synaptodendritic injury disrupts the highly integrated functioning of neural circuits that is required to process information, leading to HIV-associated neurocognitive disorders (HAND) (17–19). HAND is a spectrum of neurocognitive and degenerative conditions among which HIV-associated dementia (HAD) is the most severe form. The brain, however, does not respond passively to synaptodendritic injury but instead actively upregulates pathways promoting repair and regeneration. These pathways have become particularly important today, as survival has been prolonged by combination anti-retroviral therapy (cART). Because both viral and host factors play a role in HAND, effective treatment might require both cART as well as "adjunctive" therapies that include neuroprotective and neuro-regenerative agents.

4.3 NEUROTOXIC FACTOR OF HIV

The HIV genome codes for a variety of proteins that can damage neuronal cells and interfere with CNS function. Two of the more important viral proteins shown to be neurotoxic are gp120, the virus's envelope protein, and transactivator of transcription (Tat). gp120 is necessary for infectivity but also interacts with host cellular receptors to alter glutamate pathway signaling and induce cytokine production that can injure neurons and affect the activation state of microglia and astrocytes. Patterns of neuronal injury are seen in the autopsy brains of cognitively impaired HIV-infected individuals that have been also replicated in cultures of human nervous tissue exposed to gp120 (20) and in transgenic mice that express HIV-1 gp120 in the brain (21). These changes include synaptodendritic injury, reactive astrocytosis, and microgliosis, and loss of large pyramidal neurons (22). One molecular mechanism by which gp120 might induce these changes in the synaptodendritic circuit is through glutamate-mediated excitotoxicity, which can initiate caspase-mediated cell death or by activating intracellular metabolic signaling for neuronal autophagy (10,23).

Another viral protein reported to cause neuronal injury is Tat, which is produced by infected astrocytes (24). In experiments in which Tat-expressing astrocytes were injected into the rat dentate gyrus, Tat was taken up by granule cells and transported along neuronal pathways to the regions of the hippocampus, where it can cause glial cell activation and neurotoxicity (25). HIV Tat protein can cause mitochondrial dysfunction, dendritic loss, and cell death in neurons at concentrations lower than those needed to support viral replication (26).

4.4 SECONDARY EFFECTS POST-HIV INFECTION

Secondary effects of HIV infection on the immune system include a broad range of nervous system damage (27,28). Many of the host factors relevant to HIV CNS injury are chemical mediators of inflammation and immunity (i.e., cytokines and chemokines). Cytokines are produced by immune cells including macrophages that traffic into the CNS from the peripheral circulation, as well as by brain astrocytes and microglia when undergoing activation post-HIV infection. Chemokines represent a subset of cytokines with chemoattractant properties and are particularly important in HIV-related pathology. These chemokines signal through their cognate chemokine receptors that are expressed in the brain cell types such as microglia, astrocytes, oligodendrocytes, and neurons.

Abnormal activation of cytokine and chemokine receptors results in structural and functional changes in the neuronal environment that are minimally reversible. This suggests the scope for appropriate treatments that might promote neuronal repair and mitigate the damage arising from signaling due to abnormal activation of cytokine or chemokine receptors. Importantly, in such circumstances the severity and impact of HAND depends on the expression of these host factors, which can differ between individuals. For example, studies have shown that increased susceptibility to neurodegeneration in HIV is associated with host genetic variations that might account for differential susceptibility to HAND between different individuals infected with the same viral strain (29,30).

4.5 IMPACT OF CO-INFECTIONS DURING HIV INFECTIVITY

HIV-infected individuals frequently have comorbidities such as abuse of drugs and alcohol or infection with viral co-pathogens such as hepatitis C virus (HCV). These cofactors may contribute to CNS injury. For example, in neuropathological studies, HIV-infected individuals dying with histories of injection drug use show more activated microglia and diffuse astrogliosis in the white matter of the brain than their HIV-infected non-addicted counterparts (31–33). Experimental evidence from in vitro, in vivo, and primate models suggests that alcohol may exacerbate the immunological abnormalities associated with HIV infection, specifically the peripheral immune system (34–36). In addition, several studies have demonstrated that HCV co-infection is associated with a greater risk of neurocognitive disorders among those with HIV, with or without drug use (37,38). This could be because HCV induces systemic inflammation and possibly CNS immune activation. Thus, it is very likely that there may be some common immuno-neuropathogenic mechanisms leading to a heightened likelihood of CNS disease in co-infected individuals.

4.6 IMPACT OF HIV INFECTION ON THE SYNAPTODENDRITIC CIRCUIT

Synaptodendritic injury is a general term encompassing a variety of structural and chemical changes that ultimately can impair the normal functioning of neuronal networks. Normal synaptodendritic networks are complex and highly branched, whereas injured networks are simplified due to the loss of this architecture. Features of synaptodendritic injury include dendritic spine retraction, beading, and aberrant sprouting. Higher cognitive functions depend on the integrity of complex synaptodendritic networks, therefore damage results in deficient cognitive skills and behavior. Synaptodendritic injury is studied by immunostaining with antibodies to presynaptic synaptophysin (SYN) and postsynaptic microtubule associated protein-2 (MAP2) (39). MAP2 is expressed on neuronal cell bodies and dendrites. In HIV-infected individuals, the degree of neurocognitive impairment is strongly related to loss of immunostaining for SYN and MAP2 (40). This technique has provided evidence that the striatum and the hippocampus are particularly affected (25,40). This regional impact correlates with the higher burden of HIV proteins and viral RNA in the striatum and white matter that connects the striatum to the prefrontal cortex (41). Techniques for measuring synaptodendritic injury in living humans are still indirect and imprecise. Thus, a secondary approach is often considered. For instance, proteins released from damaged neurons into the extracellular space can be detected in the cerebrospinal fluid (CSF) or blood. Elevated CSF neurofilament protein (NFL) concentrations, for example, are thought to reflect injury to myelinated axons. The levels of NFL in the CSF are increased both in the context of HIV dementia as well as after the interruption of cART, which results in a significant rebound of HIV replication in the patients (42,43). However, it remains known whether increased NFL can result from neuronal injury alone or requires cell death. Also, the relationship between changes in neurological status in HIV patients and changes in NFL levels remains unknown. There is also gross structural atrophy in the brain that can be visualized by brain computed tomography (CT) and magnetic resonance imaging (MRI). Radiological studies have shown that white matter loss and abnormal white matter signal are closely correlated with the loss of MAP-2 immunostaining, particularly in the presence of HIV encephalitis (44). In addition, worsening of white matter abnormalities have been shown to correlate with poor cognitive impairment in HAND patients (45).

4.7 INFECTION OF ASTROCYTES

It is well established that macrophages and microglia are the major sites of HIV-1 replication in the CNS, with HIV-1–specific antigens co-localizing with microglial, multinucleated giant cells and perivascular macrophages in HIV-1-specific CNS lesions (46–48). However, the role of astrocytes in HIV-1-associated neurocognitive disorders and whether these astrocytes serve as a possible HIV-1 reservoir remains a topic of debate. Astrocytes are cells that arise from the neuroectoderm during development and are the most abundant cell population within the CNS. They have multiple biological functions, including biochemical support of brain endothelial cells essential in maintaining the BBB, production of neurotrophic factors, homeostasis of the brain microenvironment, and regulation of repair and scarring following CNS trauma. The presence of reactive astrocytosis is a significant pathological feature in HAD and chronic HIV infection. Studies have shown that a subpopulation of latently infected astrocytes undergoes apoptosis in vivo, which correlates with the severity of HAD. The mode and extent of astrocyte infection remains controversial with only trace amounts of CD4 receptor being demonstrated on the cell surface. It is well known that the HIV virion gets uptaken by the endocytic pathways and is then disseminated to susceptible cells by astrocytes (49). Alternatively, astrocytes may also be efficiently infected by cell-to-cell contact with HIV-infected lymphocytes. This may occur in the cerebral vasculature where the astrocytic foot processes may be in contact with circulating lymphocytes or in the context of immune reconstitution inflammatory syndrome when there is infiltration of lymphocytes within the parenchyma of the brain. Astrocytes have been demonstrated to contain integrated HIV using in vivo models. Recent studies using sensitive molecular techniques have demonstrated the frequency of astrocyte infection to be in the range of 15%–20%. This is indicative of a frequency of infection correlating with both the severity of HIV encephalitis and proximity to perivascular macrophages and multinucleated giant cells. Studies have shown astrocytes, in significant numbers, do contain integrated genomes, however the mechanisms underlying the reactivation and virus production from these cells remain unclear. It has been demonstrated that these cells can pass on the infectious virus possibly produced from susceptible cells in the CNS including macrophages and microglia, ultimately contributing to the spread of CNS infection. Additionally, despite evidence supporting the production of virus by astrocytes, it is well established that astrocytes can produce the HIV regulatory protein Tat. Tat-encoding mRNA has been detected in the CNS of HAD and non-demented HIV-infected individuals. Tat protein has been detected in perivascular brain macrophages in AIDS patients and latently infected astrocytes have been shown to secrete HIV-1 Tat and accumulate Tat mRNA transcripts. Tat protein is also a potent neurotoxin and has been demonstrated to cause neuronal damage in both in vitro and in vivo experimental systems. It remains unknown as to what cellular factors in astrocytes prevent viral replication despite the production of Tat protein.

4.8 NEUROLOGICAL MANIFESTATIONS IN HIV INFECTION

Complications of the nervous system can occur in more than 40% of patients with HIV. In addition, poor neurological outcomes are exacerbated in the presence of drugs of abuse such as cocaine, methamphetamine, and heroin. Research suggests the synergistic impact of drugs of abuse on molecular signaling in the brain that results in exacerbation of neurological outcomes including decline in cognitive and motor function in HIV-positive individuals (9). Aseptic meningitis and acute demyelinating polyneuropathy (AIDP) can be the some of the major presenting symptoms of acute HIV infection. Twenty-five percent of patients are estimated to develop aseptic meningitis that can occur within 2 weeks of systemic infection. In approximately 20% of cases, neurologic manifestations are the presenting signs of AIDS. Autopsy-based reports state that the prevalence of neuropathologic abnormalities is more than 50% (50). Post-mortem studies conducted early in the epidemic found pathological CNS alterations in 80% of patients (51). CT and MRI studies indicate the prevalence of progressive cerebral atrophy in AIDS, which is linked to loss of neuronal population (52–54).

4.8.1 HIV-Associated Neurocognitive Disorder

HAND defines a spectrum of neurocognitive impairments that includes asymptomatic neurocognitive impairment, mild neurocognitive disorder, and HAD (55,56). This disease spectrum is diagnosed using neuropsychological testing and functional assessments such as cognition and daily functioning in the affected individuals. As discussed above, HAD is the most severe form in this spectrum.

HAD (AIDS–dementia complex): HAD is an HIV-associated, chronic neurodegenerative clinical condition that is characterized by progressive cognitive and motor impairment and atrophic changes in the brain (57). Give the introduction of antiretroviral therapy (ART), it is true that AIDS patients have improved survival rates; however, approximately 20% of these patients develop AIDS dementia complex, and it is one of the most common causes of morbidity in this pathological group in the United States. AIDS dementia complex occurs during the later stages of AIDS and predominantly affects the white matter by inducing loss of myelin and reactive gliosis. The diagnosis is based on the clinical findings of CNS neurologic dysfunction, which include inattention, indifference, and psychomotor slowing, and on objective measures of dysfunction that include standardized neuropsychologic tests. The neurologic dysfunction can further progress to frank dementia, and the viral load in the cerebrospinal fluid helps predict the increased risk for dementia.

Several lines of evidence suggest that some HAD patients can harbor macrophage-tropic HIV-1 variants (58–62). This is important to acknowledge because it harbors a distinct phenotype associated with the ability to infect cells that have low surface expression of CD4. The initiation of ART results in rapid decay of virus in the blood and circulation, which is associated with virus replicating in activated CD4+ T cells (62,63). However, in the CSF the virus decays slowly with the initiation of therapy. This concept suggests a longer-lived infected cell type still harboring the virus (64–66). Previous studies have reported that HIV-1 populations in the CSF of HAD subjects have increased viral genetic compartmentalization compared to virus in the blood (67,68). In addition, genetically distinct HIV-1 variants have been detected at autopsy in the CNS of some subjects with HAD, suggesting the presence of an autonomous viral replication that is associated with the severity of the neurological outcomes as seen in patients (69,70).

4.9 CLINICAL CASE REPORTS

In this section we will review the various clinical presentations and neuropathological outcomes that have been observed in HIV patients that are in the spectrum of HAND. HIV infection leads to a persistent neuroinflammatory phenotype that culminates in neuronal dysfunction and cell death. As discussed in the earlier sections of this chapter, HAND pathology can include no symptoms to severe neurocognitive damage. HIV infection in the brain includes a broad range of pathologies that include mass lesions in the brain tissue, vasculopathy, as well as cerebral edema in patients across all age groups. This section includes representative clinical findings as shown in Figures 4.3–4.8 from pediatric and adult HIV patients where various forms of brain pathology have been detected by imaging-based methods.

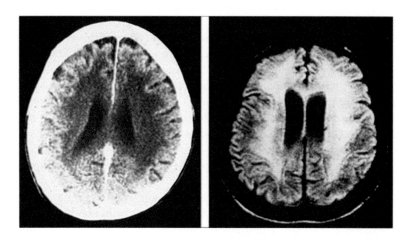

Figure 4.3 White matter abnormalities on CT and MRI. *Left:* CT scan showing ventricular enlargement and white matter hypodensity. *Right:* Fluid-attenuated inversion recovery (FLAIR)-MRI showing both cortical and central atrophy, and characteristic confluent signal abnormalities deep within the white matter.

Source: Reprinted with permission from (71).

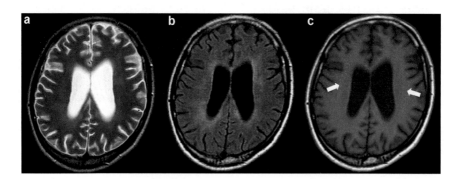

Figure 4.4 HIV encephalopathy in a 35-year-old man with HIV infection and memory loss. Brain MRI was performed a few days before the initiation of antiretroviral therapy and showed cerebral atrophy. T2-weighted (a) and FLAIR images (b) show a pale hyperintense area in the periventricular white matter. The lesion shows an iso to slightly low signal intensity on T1-weighted image (c, *white arrows*).

Source: Figure reprinted with permission from (72).

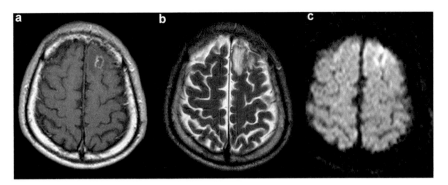

Figure 4.5 Ring-enhanced lesion in a patient with HIV-related primary CNS lymphoma. Contrast-enhanced T1-weighted image (a) shows a ring-enhanced lesion in the left frontal subcortical white matter. The mass shows a heterogeneous signal on T2-weighted image (b) with surrounding edema. On diffusion-weighted image (c), the lesion shows peripheral hyperintensity.

Source: Figure reprinted with permission from (72).

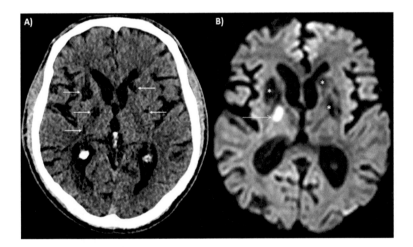

Figure 4.6 CT demonstrated multiple sequelae and lesions in the basal ganglia (*arrows*). (A) The cerebrospinal fluid analysis showed an elevated cell count (12/mm³ – N <3–5/mm³) and proteins (102 mg/dL – N <30 mg/dL). (B) The patient was hospitalized, and a brain MRI showed multiple ischemic lesions. Diffusion-weighted imaging at the same level as the CT, showing a restriction due to an ischemic event in the right basal ganglia (*arrow*) and bilateral areas of increased diffusion consistent with sequelae.

Source: Figure reprinted with permission from (73).

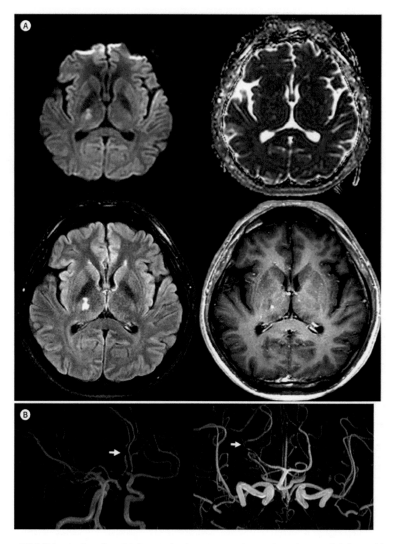

Figure 4.7 HIV-associated multiple cerebral aneurysmal vasculopathy in a 34-year-old man, presenting with left side weakness and tingling sensation in left leg and cheek. (A) MRI demonstrates subacute infarction in the right thalamus. The lesion shows high signal intensity in the diffusion-weighted image (*left upper panel*) without remarkable signal change on apparent diffusion coefficient map (*right upper panel*), high signal intensity in the FLAIR images (*left lower panel*), and subtle heterogeneous contrast enhancement (*right lower panel*). (B) 3-dimensional time-of-flight MRI reveals stenosis at right A2 (*arrow in left panel*) and P2 segment (*arrow in right panel*).

Source: Figure reprinted with permission from (74).

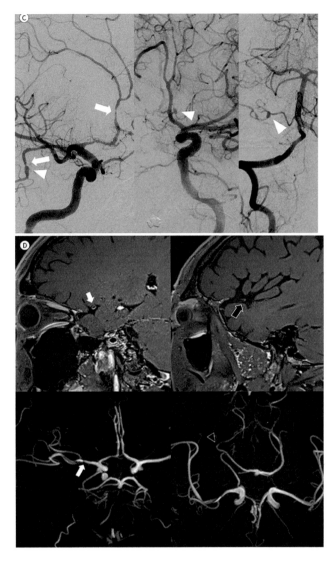

Figure 4.7 (Continued) (C) Right internal carotid artery (*left panel*), left internal carotid artery (*center panel*), and right vertebral artery (*right panel*) angiograms show multifocal stenosis (*arrows*) and aneurysmal dilatation (*arrowheads*) involving right A2, branch of A2, branch of left A2, right anterior inferior cerebellar artery, and right P2. (D) Sagittal contrast enhanced T1 plaque MRI (*upper panel*) after 3 months reveals eccentric wall thickening with enhancement of stenotic right M1 (*white arrow*) and M2 (*black arrow*). Three-dimensional time-of-flight MRI (*lower panel*) reveals new stenotic lesions in right M1 (*white arrow*) and partially improved stenotic lesions in right P2 (*black arrowhead*).

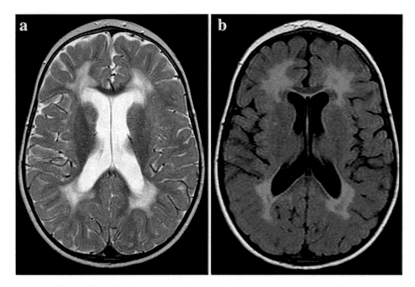

Figure 4.8 Neurological outcomes in a pediatric case of HIV infection. (a) Axial T2 and (b) FLAIR images of an HIV-positive child showing bilateral symmetrical high signal in the periventricular white matter with normal basal ganglia. Prominence of the cerebral sulci implies volume loss.

Source: Figure reprinted with permission from (75).

4.10 CONCLUSION

Several lines of experimental and clinical evidence point to the fact that the brain serves as a potent reservoir of viral particles, allowing the HIV infection to persist for a long period of time. The brain's intricate multicellular landscape adds on to the virus's infectivity cycle, thus making therapeutic intervention challenging. Significant progress has been made in understanding the cellular and molecular aspects of HIV infection, yet there are numerous questions that demand clear answers. For instance, there is a lack of understanding on CNS-specific biomarkers that could be indicative of HIV's impact in the brain. There is a need for understanding the genetic risk factors, if any, that contribute to CNS pathologies among patient populations. Therefore, it is an important subject to gain a clear understanding on the pathophysiological impact of the virus in the human brain, especially in identifying key features from experimental and clinical observations.

REFERENCES

1. Rodger AJ, Cambiano V, Bruun T, Vernazza P, Collins S, Degen O, et al. Risk of HIV transmission through condomless sex in serodifferent gay couples with the HIV-positive partner taking suppressive antiretroviral therapy (PARTNER): Final results of a multicentre, prospective, observational study.

Lancet [Internet] 2019;393(10189); 2428–2438. Available from: https://linkinghub.elsevier.com/retrieve/pii/S0140673619304180

2. Eisinger RW, Dieffenbach CW, Fauci AS. HIV viral load and transmissibility of HIV infection. *JAMA* [Internet] 2019;321(5); 451. Available from: http://jama.jamanetwork.com/article.aspx?doi=10.1001/jama.2018.21167

3. Mabuka J, Nduati R, Odem-Davis K, Peterson D, Overbaugh J. HIV-specific antibodies capable of adcc are common in breastmilk and are associated with reduced risk of transmission in women with high viral loads (Desrosiers RC, editor). *PLoS Pathogens* [Internet] 2012;8(6); e1002739. Available from: https://dx.plos.org/10.1371/journal.ppat.1002739

4. Mead MN. Contaminants in human milk: Weighing the risks against the benefits of breastfeeding. *Environmental Health Perspectives* [Internet] 2008;116(10). Available from: https://ehp.niehs.nih.gov/doi/10.1289/ehp.116-a426

5. Laurence J. T-cell subsets in health, infectious disease, and idiopathic CD4⁺T lymphocytopenia. *Annals of Internal Medicine* [Internet] 1993;119(1); 55. Available from: http://annals.org/article.aspx?doi=10.7326/0003-4819-119-1-199307010-00010

6. Joseph SB, Arrildt KT, Sturdevant CB, Swanstrom R. HIV-1 target cells in the CNS. *Journal of NeuroVirology* [Internet] 2015;21(3); 276–289. Available from: http://link.springer.com/10.1007/s13365-014-0287-x

7. Meyer AC, Njamnshi AK, Gisslen M, Price RW. Neuroimmunology of CNS HIV infection: A narrative review. *Frontiers in Neurology* [Internet] 2022; 13. Available from: www.frontiersin.org/articles/10.3389/fneur.2022.843801/full

8. Riggs PK, Chaillon A, Jiang G, Letendre SL, Tang Y, Taylor J, et al. Lessons for understanding central nervous system HIV reservoirs from the last gift program. *Current HIV/AIDS Reports* [Internet] 2022;19(6); 566–579. Available from: https://link.springer.com/10.1007/s11904-022-00628-8

9. Dash S, Balasubramaniam M, Villalta F, Dash C, Pandhare J. Impact of cocaine abuse on HIV pathogenesis. *Front Microbiol* [Internet] 2015; 6. Available from: http://journal.frontiersin.org/Article/10.3389/fmicb.2015.01111/abstract

10. Pandhare J, Dash S, Jones B, Villalta F, Dash C. A novel role of proline oxidase in HIV-1 envelope glycoprotein-induced neuronal autophagy. *Journal of Biological Chemistry* [Internet] 2015;290(42); 25439–25451. Available from: https://linkinghub.elsevier.com/retrieve/pii/S0021925820445624

11. Chivero ET, Guo M-L, Periyasamy P, Liao K, Callen SE, Buch S. HIV-1 tat primes and activates microglial NLRP3 inflammasome-mediated neuroinflammation. *Journal of Neuroscience* [Internet] 2017;37(13);3599–609. Available from: www.jneurosci.org/lookup/doi/10.1523/JNEUROSCI.3045-16.2017

12. Chatterjee N, Callen S, Seigel GM, Buch SJ. HIV-1 Tat-mediated neurotoxicity in retinal cells. *Journal of Neuroimmune Pharmacology* [Internet] 2011;6(3); 399–408. Available from: http://link.springer.com/10.1007/s11481-011-9257-8

13. Niu F, Liao K, Hu G, Moidunny S, Roy S, Buch S. HIV tat-mediated induction of monocyte transmigration across the blood–brain barrier: Role of chemokine receptor CXCR3. *Frontiers in Cell and Developmental Biology* [Internet] 2021; 9. Available from: www.frontiersin.org/articles/10.3389/fcell.2021.724970/full

14. Torres L, Noel RJ. Astrocytic expression of HIV-1 viral protein R in the hippocampus causes chromatolysis, synaptic loss and memory impairment. *Journal of Neuroinflammation* [Internet] 2014;11(1); 53. Available from: https://jneuroinflammation.biomedcentral.com/articles/10.1186/1742-2094-11-53

15. Fischer-Smith T, Rappaport J. Evolving paradigms in the pathogenesis of HIV-1-associated dementia. *Expert Reviews in Molecular Medicine* [Internet] 2005;7(27). Available from: www.journals.cambridge.org/abstract_S1462399405010239

16. Rao VR, Ruiz AP, Prasad VR. Viral and cellular factors underlying neuropathogenesis in HIV associated neurocognitive disorders (HAND). *AIDS Research and Therapy* [Internet] 2014;11(1); 13. Available from: https://aidsrestherapy.biomedcentral.com/articles/10.1186/1742-6405-11-13

17. Nash B, Festa L, Lin C, Meucci O. Opioid and chemokine regulation of cortical synaptodendritic damage in HIV-associated neurocognitive disorders. *Brain Research* [Internet] 2019;1723; 146409. Available from: https://linkinghub.elsevier.com/retrieve/pii/S0006899319304639

18. Irollo E, Luchetta J, Ho C, Nash B, Meucci O. Mechanisms of neuronal dysfunction in HIV-associated neurocognitive disorders. *Cellular and Molecular Life Sciences* [Internet] 2021;78(9); 4283–4303. Available from: https://link.springer.com/10.1007/s00018-021-03785-y

19. Ru W, Tang SJ. HIV-associated synaptic degeneration. *Molecular Brain* [Internet] 2017;10(1); 40. Available from: https://molecularbrain.biomedcentral.com/articles/10.1186/s13041-017-0321-z

20. Iskander S, Walsh KA, Hammond RR. Human CNS cultures exposed to HIV-1 gp120 reproduce dendritic injuries of HIV-1-associated dementia. *Journal of Neuroinflammation* [Internet] 2004;1(1); 7. Available from: www.ncbi.nlm.nih.gov/pubmed/15285795

21. Toggas SM, Masliah E, Rockenstein EM, Rall GF, Abraham CR, Mucke L. Central nervous system damage produced by expression of the HIV-1 coat protein gp120 in transgenic mice. *Nature* [Internet] 1994;367(6459); 188–193. Available from: www.ncbi.nlm.nih.gov/pubmed/8114918

22. Kaul M, Lipton SA. Experimental and potential future therapeutic approaches for HIV-1 associated dementia targeting receptors for chemokines, glutamate and erythropoietin. *Neurotoxicity Research* [Internet] 2005;8(1–2); 167–186. Available from: http://link.springer.com/10.1007/BF03033828

23. Tenneti L, Lipton SA. Involvement of activated caspase-3-like proteases in N-methyl-D-aspartate-induced apoptosis in cerebrocortical neurons. *Journal of Neurochemistry* [Internet] 2001;74(1); 134–142. Available from: http://doi.wiley.com/10.1046/j.1471-4159.2000.0740134.x

24. Nath A. Human immunodeficiency virus (HIV) proteins in neuropathogenesis of HIV dementia. *Journal of Infectious Diseases* [Internet] 2002;186(s2); S193–S198. Available from: https://academic.oup.com/jid/article-lookup/doi/10.1086/344528

25. Bruce-Keller AJ, Chauhan A, Dimayuga FO, Gee J, Keller JN, Nath A. Synaptic transport of human immunodeficiency virus-Tat protein causes neurotoxicity and gliosis in rat brain. *Journal of Neuroscience* [Internet] 2003;23(23); 8417–8422. Available from: www.ncbi.nlm.nih.gov/pubmed/12968004

26. Chauhan A, Turchan J, Pocernich C, Bruce-Keller A, Roth S, Butterfield DA, et al. Intracellular human immunodeficiency virus Tat expression in astrocytes promotes astrocyte survival but induces potent neurotoxicity at distant sites via axonal transport. *Journal of Biological Chemistry* [Internet] 2003;278(15); 13512–13159. Available from: https://linkinghub.elsevier.com/retrieve/pii/S0021925819647368

27. Deeks SG, Overbaugh J, Phillips A, Buchbinder S. HIV infection. *Nature Reviews Disease Primers* [Internet] 2015;1(1); 15035. Available from: www.nature.com/articles/nrdp201535

28. Kranick SM, Nath A. Neurologic complications of HIV-1 infection and its treatment in the era of antiretroviral therapy. *Continuum: Lifelong Learning in Neurology* [Internet] 2012;18(6); 1319–1337. Available from: http://journals.lww.com/00132979-201212000-00010

29. Gonzalez E, Rovin BH, Sen L, Cooke G, Dhanda R, Mummidi S, et al. HIV-1 infection and AIDS dementia are influenced by a mutant MCP-1 allele linked to increased monocyte infiltration of tissues and MCP-1 levels. *Proceedings of the National Academy of Sciences* [Internet] 2002;99(21); 13795–13800. Available from: https://pnas.org/doi/full/10.1073/pnas.202357499

30. Quasney MW, Zhang Q, Sargent S, Mynatt M, Glass J, McArthur J. Increased frequency of the tumor necrosis factor-α-308 a allele in adults with human immunodeficiency virus dementia. *Annals of Neurology* [Internet] 2001;50(2); 157–162. Available from: https://onlinelibrary.wiley.com/doi/10.1002/ana.1284

31. Cook JE, Dasgupta S, Middaugh LD, Terry EC, Gorry PR, Wesselingh SL, et al. Highly active antiretroviral therapy and human immunodeficiency virus encephalitis. *Annals of Neurology* [Internet] 2005;57(6); 795–803. Available from: https://onlinelibrary.wiley.com/doi/10.1002/ana.20479

32. Persidsky Y, Limoges J, McComb R, Bock P, Baldwin T, Tyor W, et al. Human immunodeficiency virus encephalitis in SCID mice. *American Journal of Pathology* [Internet] 1996;149(3); 1027–1053. Available from: www.ncbi.nlm.nih.gov/pubmed/8780406

33. Langford TD, Letendre SL, Larrea GJ, Masliah E. Changing patterns in the neuropathogenesis of HIV during the HAART era. *Brain Pathology* [Internet] 2003;13(2); 195–210. Available from: www.ncbi.nlm.nih.gov/pubmed/12744473

34. Bagasra O, Balla AK, Lischner HW, Pomerantz RJ. Alcohol intake increases human immunodeficiency virus type 1 replication in human peripheral blood mononuclear cells. *Journal of Infectious Diseases* [Internet] 1993;167(4); 789–797. Available from: https://academic.oup.com/jid/article-lookup/doi/10.1093/infdis/167.4.789

35. Tyor WR, Middaugh LD. Do alcohol and cocaine abuse alter the course of HIV-associated dementia complex? *Journal of Leukocyte Biology* [Internet] 1999;65(4); 475–481. Available from: https://onlinelibrary.wiley.com/doi/10.1002/jlb.65.4.475

36. Winsauer PJ, Moerschbaecher JM, Brauner IN, Purcell JE, Lancaster JR, Bagby GJ, et al. Alcohol unmasks simian immunodeficiency virus-induced cognitive impairments in rhesus monkeys. *Alcoholism: Clinical and Experimental Research* [Internet] 2002;26(12); 1846–1857. Available from: www.ncbi.nlm.nih.gov/pubmed/12500109

37. Cherner M, Ellis RJ, Lazzaretto D, Young C, Mindt MR, Atkinson JH, et al. Effects of HIV-1 infection and aging on neurobehavioral functioning: preliminary findings. *AIDS* [Internet] 2004;18 (Suppl 1); S27–S34. Available from: www.ncbi.nlm.nih.gov/pubmed/15075495

38. Morgello S, Estanislao L, Ryan E, Gerits P, Simpson D, Verma S, et al. Effects of hepatic function and hepatitis C virus on the nervous system assessment of advanced-stage HIV-infected individuals. *AIDS* [Internet] 2005;19 (Suppl 3); S116–S122. Available from: www.ncbi.nlm.nih.gov/pubmed/16251806

39. Orenstein JM, Meltzer MS, Phipps T, Gendelman HE. Cytoplasmic assembly and accumulation of human immunodeficiency virus types 1 and 2 in recombinant human colony-stimulating factor-1-treated human monocytes: An ultrastructural study. *Journal of Virology* [Internet] 1988;62(8); 2578–2786. Available from: www.ncbi.nlm.nih.gov/pubmed/3260631

40. Moore DJ, Masliah E, Rippeth JD, Gonzalez R, Carey CL, Cherner M, et al. Cortical and subcortical neurodegeneration is associated with HIV neurocognitive impairment. *AIDS* [Internet] 2006;20(6); 879–887. Available from: www.ncbi.nlm.nih.gov/pubmed/16549972

41. Masliah E, Ellis RJ, Mallory M, Heaton RK, Marcotte TD, Nelson JA, et al. Dendritic injury is a pathological substrate for human immunodeficiency virus? Related cognitive disorders. *Annals of Neurology* [Internet] 1997;42(6); 963–972. Available from: https://onlinelibrary.wiley.com/doi/10.1002/ana.410420618

42. Arnold SE. Contributions of neuropathology to understanding schizophrenia in late life. *Harvard Review of Psychiatry* [Internet] 2001;9(2); 69–76. Available from: http://informahealthcare.com/doi/abs/10.1080/10673220127882

43. Law AJ. Reduced spinophilin but not microtubule-associated protein 2 expression in the hippocampal formation in schizophrenia and mood disorders: Molecular evidence for a pathology of dendritic spines. *American Journal of Psychiatry* [Internet] 2004;161(10); 1848–1855. Available from: http://ajp.psychiatryonline.org/article.aspx?articleID=177100

44. Archibald SL, Masliah E, Fennema-Notestine C, Marcotte TD, Ellis RJ, McCutchan JA, et al. Correlation of in vivo neuroimaging abnormalities with postmortem human immunodeficiency virus encephalitis and dendritic loss. *Archives of Neurology* [Internet] 2004;61(3); 369. Available from: http://archneur.jamanetwork.com/article.aspx?doi=10.1001/archneur.61.3.369

45. Everall I. Lithium ameliorates HIV-gp120-Mediated neurotoxicity. *Molecular and Cellular Neuroscience* [Internet] 2002;21(3); 493–501. Available from: https://linkinghub.elsevier.com/retrieve/pii/S1044743102911966

46. Wallet C, De Rovere M, Van Assche J, Daouad F, De Wit S, Gautier V, et al. Microglial cells: The main HIV-1 reservoir in the Brain. *Frontiers in Cellular and Infection Microbiology* [Internet] 2019; 9. Available from: www.frontiersin.org/article/10.3389/fcimb.2019.00362/full

47. Koppensteiner H, Brack-Werner R, Schindler M. Macrophages and their relevance in human immunodeficiency virus type i infection. *Retrovirology* [Internet] 2012;9(1); 82. Available from: https://retrovirology.biomedcentral.com/articles/10.1186/1742-4690-9-82

48. Hendricks CM, Cordeiro T, Gomes AP, Stevenson M. The interplay of HIV-1 and macrophages in viral persistence. *Frontiers in Microbiology* [Internet] 2021; 12. Available from: www.frontiersin.org/articles/10.3389/fmicb.2021.646447/full

49. Valdebenito S, Castellano P, Ajasin D, Eugenin EA. Astrocytes are HIV reservoirs in the brain: A cell type with poor HIV infectivity and replication but efficient cell-to-cell viral transfer. *Journal of Neurochemistry* [Internet] 2021;158(2); 429–443. Available from: https://onlinelibrary.wiley.com/doi/10.1111/jnc.15336

50. Debalkie Animut M, Sorrie MB, Birhanu YW, Teshale MY. High prevalence of neurocognitive disorders observed among adult people living with HIV/AIDS in Southern Ethiopia: A cross-sectional study (Cysique LA, editor). *PLoS One* [Internet] 2019;14(3); e0204636. Available from: https://dx.plos.org/10.1371/journal.pone.0204636

51. Navia BA, Jordan BD, Price RW. The AIDS dementia complex: I. Clinical features. *Annals of Neurology* [Internet] 1986;19(6); 517–524. Available from: https://onlinelibrary.wiley.com/doi/10.1002/ana.410190602

52. Nir TM, Jahanshad N, Ching CRK, Cohen RA, Harezlak J, Schifitto G, et al. Progressive brain atrophy in chronically infected and treated HIV+ individuals. *Journal of NeuroVirology* [Internet] 2019;25(3); 342–353. Available from: http://link.springer.com/10.1007/s13365-019-00723-4

53. Bairwa D, Kumar V, Vyas S, Das BK, Srivastava AK, Pandey RM, et al. Case control study: Magnetic resonance spectroscopy of brain in HIV infected patients. *BMC Neurology* [Internet] 2016;16(1); 99. Available from: http://bmcneurol.biomedcentral.com/articles/10.1186/s12883-016-0628-x

54. Thompson PM, Jahanshad N. Novel neuroimaging methods to understand How HIV affects the brain. *Current HIV/AIDS Reports* [Internet] 2015;12(2); 289–298. Available from: http://link. springer.com/10.1007/s11904-015-0268-6

55. Saylor D, Dickens AM, Sacktor N, Haughey N, Slusher B, Pletnikov M, et al. HIV-associated neurocognitive disorder—Pathogenesis and prospects for treatment. *Nature Reviews Neurology* [Internet] 2016;12(4); 234–248. Available from: www.nature.com/articles/nrneurol.2016.27

56. Clifford DB, Ances BM. HIV-associated neurocognitive disorder. *Lancet Infectious Diseases* [Internet] 2013;13(11); 976–986. Available from: https://linkinghub.elsevier.com/retrieve/pii/S147 330991370269X

57. Smith AB, Smirniotopoulos JG, Rushing EJ. Central nervous system infections associated with human immunodeficiency virus infection: Radiologic-pathologic correlation. *RadioGraphics* [Internet] 2008;28(7); 2033–2058. Available from: http://pubs.rsna.org/doi/10.1148/rg.287085135

58. Gorry PR, Bristol G, Zack JA, Ritola K, Swanstrom R, Birch CJ, et al. Macrophage tropism of human immunodeficiency virus type 1 isolates from brain and lymphoid tissues predicts neurotropism independent of coreceptor specificity. *Journal of Virology* [Internet] 2001;75(21); 10073–10089. Available from: https://journals.asm.org/doi/10.1128/JVI.75.21.10073-10089.2001

59. Koenig S, Gendelman HE, Orenstein JM, Dal Canto MC, Pezeshkpour GH, Yungbluth M, et al. Detection of AIDS virus in macrophages in brain tissue from AIDS patients with encephalopathy. *Science* (80-) [Internet] 1986;233(4768); 1089–1093. Available from: www.science.org/doi/10.1126/ science.3016903

60. Rossi F, Querido B, Nimmagadda M, Cocklin S, Navas-Martín S, Martín-García J. The V1-V3 region of a brain-derived HIV-1 envelope glycoprotein determines macrophage tropism, low CD4 dependence, increased fusogenicity and altered sensitivity to entry inhibitors. *Retrovirology* [Internet] 2008;5(1); 89. Available from: https://retrovirology.biomedcentral.com/articles/10.1186/1742-4690-5-89

61. Peters PJ, Bhattacharya J, Hibbitts S, Dittmar MT, Simmons G, Bell J, et al. Biological analysis of human immunodeficiency virus type 1 R5 envelopes amplified from brain and lymph node tissues of AIDS patients with neuropathology reveals two distinct tropism phenotypes and identifies envelopes in the brain that confer an enhanced. *Journal of Virology* [Internet] 2005;79(5); 3227–3227. Available from: https://journals.asm.org/doi/10.1128/JVI.79.5.3227.2005

62. Ho DD, Neumann AU, Perelson AS, Chen W, Leonard JM, Markowitz M. Rapid turnover of plasma virions and CD4 lymphocytes in HIV-1 infection. *Nature* [Internet] 1995;373(6510); 123–126. Available from: www.nature.com/articles/373123a0

63. Wei X, Ghosh SK, Taylor ME, Johnson VA, Emini EA, Deutsch P, et al. Viral dynamics in human immunodeficiency virus type 1 infection. *Nature* [Internet] 1995;373(6510); 117–122. Available from: www.nature.com/articles/373117a0

64. Haas DW, Clough LA, Johnson BW, Harris VL, Spearman P, Wilkinson GR, et al. Evidence of a source of HIV type 1 within the central nervous system by ultraintensive sampling of cerebrospinal fluid and plasma. *AIDS Research and Human Retroviruses* [Internet] 2000;16(15); 1491–1502. Available from: www.liebertpub.com/doi/10.1089/088922200750006010

65. Ellis RJ, Gamst AC, Capparelli E, Spector SA, Hsia K, Wolfson T, et al. Cerebrospinal fluid HIV RNA originates from both local CNS and systemic sources. *Neurology* [Internet] 2000;54(4); 927–936. Available from: www.neurology.org/lookup/doi/10.1212/WNL.54.4.927

66. Schnell G, Spudich S, Harrington P, Price RW, Swanstrom R. Compartmentalized human immunodeficiency virus type 1 originates from long-lived cells in some subjects with HIV-1–Associated dementia (Douek DC, editor). *PLoS Pathogens* [Internet] 2009;5(4); e1000395. Available from: https://dx.plos.org/10.1371/journal.ppat.1000395

67. Harrington PR, Schnell G, Letendre SL, Ritola K, Robertson K, Hall C, et al. Cross-sectional characterization of HIV-1 env compartmentalization in cerebrospinal fluid over the full disease course. *AIDS* [Internet] 2009;23(8); 907–915. Available from: https://journals.lww.com/00002030-200905150-00005

68. Pillai SK, Pond SLK, Liu Y, Good BM, Strain MC, Ellis RJ, et al. Genetic attributes of cerebrospinal fluid-derived HIV-1 env. *Brain* [Internet] 2006;129(7); 1872–1883. Available from: https://academic.oup.com/brain/article-lookup/doi/10.1093/brain/awl136

69. Dunfee RL, Thomas ER, Gorry PR, Wang J, Taylor J, Kunstman K, et al. The HIV Env variant N283 enhances macrophage tropism and is associated with brain infection and dementia. *Proceedings of the National Academy of Sciences of the United States of America* [Internet] 2006;103(41); 15160–15165. Available from: https://pnas.org/doi/full/10.1073/pnas.0605513103

70. Power C, McArthur JC, Johnson RT, Griffin DE, Glass JD, Perryman S, et al. Demented and nondemented patients with AIDS differ in brain-derived human immunodeficiency virus type 1 envelope

sequences. *Journal of Virology* [Internet] 1994;68(7); 4643–4649. Available from: https://journals.asm.org/doi/10.1128/jvi.68.7.4643-4649.1994

71. McArthur JC, Brew BJ, Nath A. Neurological complications of HIV infection. *Lancet Neurology* [Internet] 2005;4(9); 543–555. Available from: https://linkinghub.elsevier.com/retrieve/pii/S1474442205701654

72. Sakai M, Higashi M, Fujiwara T, Uehira T, Shirasaka T, Nakanishi K, et al. MRI imaging features of HIV-related central nervous system diseases: Diagnosis by pattern recognition in daily practice. *Japanese Journal of Radiology* [Internet] 2021;39(11); 1023–1038. Available from: https://link.springer.com/10.1007/s11604-021-01150-4

73. Arktout S. Vessel wall MRI in HIV-Associated cerebral angiitis. *Journal of the Belgian Society of Radiology* [Internet] 2020;104(1). Available from: http://jbsr.be/articles/10.5334/jbsr.2162/

74. Kim D, Baek JW, Heo YJ, Jeong HW. Human immunodeficiency virus-associated multiple cerebral aneurysmal vasculopathy in a young adult: A case report. *Journal of the Korean Society of Radiology* [Internet]. 2019;80(1); 157. Available from: https://jksronline.org/DOIx.php?id=10.3348/jksr.2019.80.1.157

75. Donald KA, Walker KG, Kilborn T, Carrara H, Langerak NG, Eley B, et al. HIV encephalopathy: Pediatric case series description and insights from the clinic coalface. *AIDS Research and Therapy* [Internet] 2015;12(1); 2. Available from: www.aidsrestherapy.com/content/12/1/2

Chapter 5

Neuropathological Complications of Zika Virus

Reshma Bhagat

5.1 INTRODUCTION

The Zika virus (ZIKV) is a mosquito-borne virus belonging to the *Flaviviri-dae* family and the *Flavivirus* genus. It was first identified in 1947, and for many years it was considered a relatively mild pathogen causing only mild symptoms in infected individuals. However, in recent years ZIKV has gained significant attention and become a major international public health concern due to large outbreaks and its association with certain complications. The active transmission of ZIKV has been reported in numerous countries worldwide, with the World Health Organization (WHO) reporting active transmission in 84 countries as of their 2017 Situation Report. ZIKV is primarily transmitted through the bites of infected *Aedes* mosquitoes, particularly *Aedes aegypti* and *Aedes albopictus* (WHO 2017; Pielnaa et al. 2020). These mosquitoes are commonly found in tropical and subtropical regions. ZIKV can also be transmitted through other routes such as sexual transmission, blood transfusion, and from mother to child during pregnancy or childbirth (Antoniou et al. 2020; Foy et al. 2011; Musso et al. 2015; Prisant et al. 2016). ZIKV infections often result in mild symptoms such as fever, rash, joint pain, and conjunctivitis, the major concern arises from its potential association with neurological complications, especially in pregnant women. ZIKV infection during pregnancy can lead to congenital Zika syndrome, resulting in severe brain abnormalities in infants (Figure 5.1). ZIKV is the only flavivirus that has teratogenic effects in humans, including microcephaly, intracranial calcification, and fetal death (van der Linden et al. 2016; MacNamara 1954).

ZIKV is a single-stranded, positive-sense RNA virus. Its genome is approximately 10.8 kilobases (kb) in size. The genome consists of a single open reading frame (ORF) of around 10 kb, with approximately 100 nucleotides of 5′ untranslated region (UTR) and 420 nucleotides of 3′ UTR (Kuno and Chang 2007). The ORF of the ZIKV genome encodes a polyprotein, which is later processed to produce the structural and non-structural proteins of the virus (Fernandez-Garcia et al. 2009). The structural proteins include the precursor membrane protein (prM), the envelope protein (E), and the capsid protein (C). The non-structural proteins consist of NS1, NS2A, NS2B, NS3, NS4A, NS4B, and NS5. The structural proteins prM, C, and E are responsible for forming the viral particle. The capsid protein (C) assembles to create the viral icosahedral capsid, while the membrane (M) and envelope (E)

DOI: 10.1201/9781003285823-5

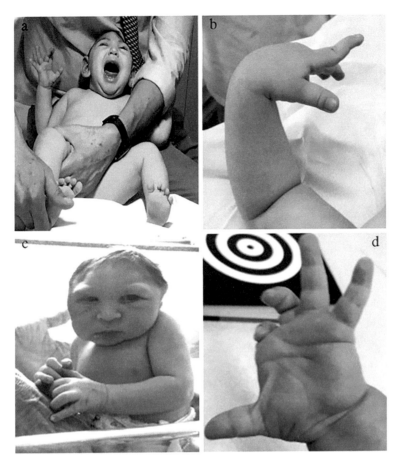

Figure 5.1 Neurological examination in congenital Zika syndrome (CZS). Common findings in the neurological examination of infants with CZS. (a) Irritability with constant inconsolable cry. (b) Swan neck position of the second finger caused by dystonic hyperextension. (c) Marked hypertonia leading to an almost stable sitting position right after birth. (d) Dystonic position of the fingers, with flexion of 2–3 and extension of 3–4.

Source: Figure reprinted with permission from Mulkey et al. (2020).

proteins are transmembrane proteins that anchor onto the outer membrane of the host cell. The envelope protein (E) plays a crucial role in host cell binding and membrane fusion (Pierson and Kielian 2013) and host response (Bhagat et al. 2018, 2021a; Kaur et al. 2023). The non-structural proteins (NS1, NS2A, NS2B, NS3, NS4A, NS4B, and NS5) are essential for various viral functions, including genome replication, polyprotein processing, and manipulation of host responses to benefit the virus. The genetic organization of ZIKV is depicted in Figure 5.2.

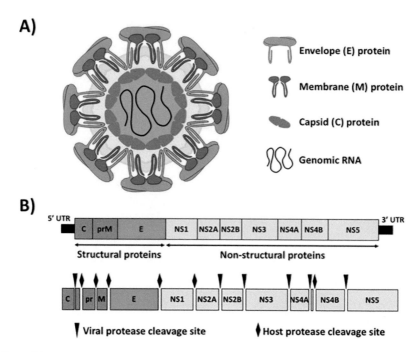

Figure 5.2 ZIKV virion structure and genetic organization. (A) Schematic representation of ZIKV virion structure: ZIKV virion surface is decorated with the E and M proteins, anchored in a lipid bilayer with an icosahedral-like symmetry. (B) Genome organization and polyprotein processing: ZIKV genome is translated as a single polyprotein that is cleaved and post-translationally by viral (*arrows*) and host (*diamonds*) proteases to yield the three structural proteins C, M, and E; and seven NS proteins NS1, NS2A, NS2B, NS3, NS4A, NS4B, and NS5.

Source: Figure reprinted with permission from Ávila-Pérez et al. (2018).

ZIKV was first isolated in 1947 from a sentinel rhesus monkey in the Zika Forest of Uganda. The isolation of the virus was part of a study supported by the Rockefeller Foundation, which aimed to investigate the enzootic or sylvatic cycle of the yellow fever virus (YFV) and identify other arboviruses. In the study conducted by Dick et al. in 1952, six sentinel platforms were set up in the Zika Forest, each containing caged rhesus monkeys (Dick 1952). One of the monkeys in Cage No. 766 exhibited fever symptoms. On the third day of fever, a blood sample was collected from this monkey. The collected blood sample was then used for intracerebral and intraperitoneal inoculation into Swiss mice, as well as subcutaneous inoculation into another rhesus monkey in Cage No. 771. The mice that were inoculated intracerebrally showed signs of sickness on the 10th day after the inoculation, indicating the presence of an infectious agent. However, the rhesus monkey in Cage No. 771 did not show any signs of illness. The infectious agent isolated from the blood sample of monkey No. 766 was subsequently referred to as the Zika virus (ZIKV), specifically the ZIKV 766 strain, which is considered the Asian strain of the virus. This discovery marked the first identification and isolation of the ZIKV, laying the foundation for further research and understanding of its characteristics, transmission, and associated health effects. In 1954, the first human isolate of ZIKV was obtained from a 10-year-old Nigerian female (MacNamara 1954). This case marked the initial confirmation of ZIKV infection in humans. Outside of Africa, ZIKV was isolated from *Aedes aegypti* mosquitoes in 1969 in Malaysia (Marchette et al. 1969).

The ZIKV is primarily recognized for its association with neurological complications, particularly when infection occurs during pregnancy. The most significant neuropathological complication linked to ZIKV infection is congenital Zika syndrome (CZS), which arises when a pregnant woman becomes infected and transmits the virus to her fetus. CZS can result in severe brain abnormalities in newborns, including microcephaly (abnormally small head size) and other developmental defects. These brain abnormalities can give rise to various neurological issues, such as intellectual disabilities, seizures, impaired vision and hearing, and motor impairments. In addition to CZS, ZIKV infection in adults can occasionally lead to neurological complications, albeit they are relatively uncommon. One such complication is Guillain–Barré syndrome (GBS), an autoimmune disorder affecting the peripheral nervous system. GBS can cause muscle weakness, paralysis, and, in severe cases, respiratory failure. It is important to note that not all individuals infected with ZIKV will experience neurological complications. The majority of ZIKV infections are mild or show no symptoms, and severe complications are infrequent. However, pregnant women and their fetuses face the highest risk of developing Zika-related neurological complications.

5.2 PERIPHERAL AND CENTRAL NERVOUS SYSTEM COMPLICATIONS IN ZIKV INFECTION

5.2.1 Zika Fever

The incubation period of flaviviruses, including ZIKV, refers to the time between being bitten by an infected mosquito and the appearance of symptoms. In the case of ZIKV, the incubation period typically ranges from 2 to 7 days (Duffy et al. 2009; Olson et al. 1981). These symptoms can include fever, rash, joint pain, muscle aches, headache, and conjunctivitis. Unlike some other flavivirus infections such as dengue fever and chikungunya fever, the onset of symptoms in Zika fever is not as clearly defined (Gourinat et al. 2015; Musso et al. 2015). There is no sudden or abrupt clinical onset that can be easily pinpointed. This can make it challenging to determine precisely when symptoms first appear in individuals with Zika fever. The illness caused by ZIKV infection is generally considered mild and self-limiting. The symptoms typically last for a few days to a few weeks in adults. It is important to note that the severity and duration of symptoms can vary from person to person. One notable characteristic of Zika fever is the absence of reported hemorrhagic events, which distinguishes it from other flavivirus infections such as dengue fever. Hemorrhagic events, which involve bleeding complications, are commonly associated with diseases like dengue fever but have not been reported in relation to Zika fever.

5.2.2 Guillain–Barré Syndrome

GBS is an acute, self-limited polyneuropathy (Donofrio 2017). It is a demyelinating disease of the peripheral nervous system also referred to as acute inflammatory demyelinating polyneuropathy. It represents a range of symptoms from a very mild case with a brief weakness to devastating paralysis, leaving the person unable to breathe independently. AIDP, acute motor axonal neuropathy (AMAN), and acute motor sensory axonal neuropathy (AMSAN) are considered clinical variants of GBS. These variants are primarily defined through electrophysiological studies, which are further supported by pathological findings. In AIDP, the most common variant of GBS, there is inflammation and damage to the myelin sheath of peripheral nerves, leading to a loss of nerve conduction and subsequent muscle weakness. This demyelination is observed in nerve conduction studies and is associated with the characteristic clinical features of GBS. AMAN and AMSAN, on the other hand, are characterized by damage primarily to the axons of peripheral nerves, resulting in motor deficits. These variants are more commonly associated with *Campylobacter jejuni* infections and are often observed in certain geographical regions. The underlying cause of GBS is believed to be due to a loss of immunological tolerance to self-antigens (Shoenfeld et al. 1996). There is evidence that the antibodies bind on the outer myelin surface producing complement activation and myelin destruction (Hafer-Macko et al. 1996). GBS has also been observed in other arboviral infections, such as dengue virus and chikungunya virus but is a rare event. ZIKV-induced GBS has been transient in duration, and most patients fully recover. No deaths were reported during the French Polynesian outbreak due to GBS (Wielanek et al. 2007).

5.2.3 Congenital Zika Syndrome

In a study of pregnant women with ZIKV infection, abnormally high rates of early miscarriages and intrauterine fetal demise were reported. In about 20% of ZIKV-infected pregnancies, compromised fetal growth and placental dysfunction have also been reported (Buathong et al. 2015; Sarno et al. 2016). CZS is characterized by a distinct pattern of structural anomalies and functional disabilities in both the central and peripheral nervous systems. One of the hallmark features of CZS is severe microcephaly, which refers to a head circumference below the third percentile for sex and age. Studies have reported a high risk of severe microcephaly, ranging from 1% to 15%, in confirmed cases of prenatal ZIKV infections.

Imaging studies have provided further insights into the neurological abnormalities associated with CZS. These include cortical disorders, ventriculomegaly (enlargement of the brain's ventricles), lissencephaly (smooth brain surface), cerebral calcifications (calcium deposits in the brain), and calcifications in the periventricular, cortical, and subcortical areas (Figure 5.3). Other observed anomalies include cortical development errors, hydrocephaly (accumulation of cerebrospinal fluid in the brain), and pachygyria/agyria (abnormal brain folding) (Antoniou et al. 2020; Nithiyanantham and Badawi 2019).

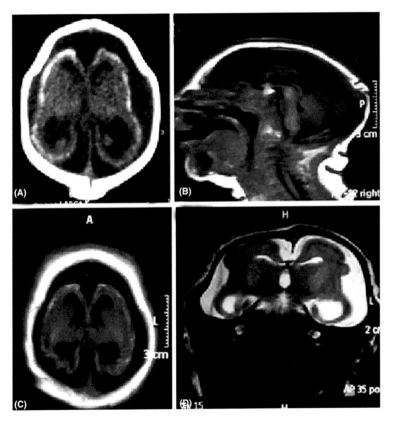

Figure 5.3 Severe microcephaly, cerebral atrophy with ventriculomegaly, and prominent cerebrospinal fluid space. Axial CT scan (A) shows extensive, punctate cortico-subcortical calcifications, cerebral atrophy, and bone overlap. Sagittal T1 MR image (B) shows craniofacial disproportion, occipital protuberance with redundant posterior skinfolds, and a hypoplastic corpus callosum. The fossa is relatively preserved. Axial T1 (C) and coronal T2 (D) demonstrate the abnormal gyral pattern with diffuse undersulcated cortex.

Source: Figure reprinted with permission from Coelho et al. (2018).

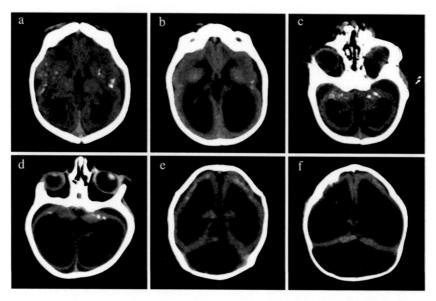

Figure 5.4 Changes in brain imaging observed during the first year of life. CT of the brain of the patient in the first month of life in the upper row in (a, c, e), at 6 months of age in lower rows (b, d), and at 13 months of age in (f). Calcifications are more difficult to visualize at 6 months of age (b, d) than at birth (a, b). There is increase in volume of intraventricular and extra-axial fluids (*bottom images*) in comparison to the initial exam (*top images*). The increasing ventricular dilatation did not show clinical or imaging signs of increased intracranial pressure at 6 months (b, d), but a high tentorium confirmed progressive hydrocephalus requiring surgical shunting at 13 months (f).

Source: Figure reprinted with permission from Schuler-Faccini et al. (2022).

Craniofacial disproportion, which refers to an imbalance in the size and shape of the skull and face, is also commonly observed in newborns affected by CZS (van der Linden et al. 2016). It is important to note that the neurological anomalies associated with CZS may continue to manifest in affected newborns as they grow, even if their head circumference falls within the normal range (Figure 5.4). This suggests that the impact of ZIKV infection on the nervous system can extend beyond the initial period of infancy (van der Linden et al. 2016).

In a clinical study (van der Linden et al. 2016) involving 141 children with congenital ZIKV infection, neuroimaging studies were conducted to assess the structural brain abnormalities present in these infants. The results revealed that all the children had detectable structural brain abnormalities, indicating the impact of ZIKV on brain development. Furthermore, the study examined the prevalence of epilepsy among these children.

The findings showed a high incidence of epilepsy, with 67% of the children experiencing seizures. The average age at which epilepsy onset occurred was 4.9 months. It was observed that the majority of infants (74%) started experiencing seizures within the first 6 months of their lives. The specific types of seizures observed were epileptic spasms (72% of cases), focal motor seizures (21% of cases), and tonic seizures (4% of cases). In another follow-up study (Satterfield-Nash Ashley et al. 2017) conducted by the National Center on Birth Defects and Developmental Disabilities at the Centers for Disease Control and Prevention, the focus was on 19 children who were born with severe microcephaly and tested positive for ZIKV infection. These children, at an age range of 19–24 months post-ZIKV infection, exhibited severe functional limitations. These limitations included motor impairment, seizure disorders, hearing and vision abnormalities, and sleep difficulties. These findings emphasize the serious neurological consequences of congenital ZIKV infection. The high prevalence of epilepsy and the presence of severe functional limitations indicate the long-term impact on the affected children's development and quality of life. It highlights the need for ongoing research, support, and intervention to address the challenges faced by these children and provide appropriate care and management for their specific needs.

BIBLIOGRAPHY

Acosta-Ampudia, Yeny, et al. 2018, Apr 11. "Autoimmune neurological conditions associated with Zika virus infection." *Frontiers in Molecular Neuroscience* 11: 116.

Antoniou, Evangelia, et al. 2020. "Zika virus and the risk of developing microcephaly in infants: A systematic review." *International Journal of Environmental Research and Public Health* 17(11): 3806.

Bhagat, Reshma, et al. 2018. "Zika virus E protein alters the properties of human fetal neural stem cells by modulating microrna circuitry." *Cell Death and Differentiation* 25(10): 1837–1854.

Bhagat, Reshma, et al. 2021a, Nov. Zika virus E protein dysregulate Mir-204/WNT2 signalling in human fetal neural stem cells. *Brain Research Bulletin* 176: 93–102.

Bhagat, Reshma, Guneet Kaur, and Pankaj Seth. 2021b, Sep. "Molecular mechanisms of Zika virus pathogenesis: An update." *Indian Journal of Medical Research* 154(3): 433–445.

Buathong, Rome, et al. 2015, Aug 5. "Detection of Zika virus infection in Thailand, 2012–2014." *American Journal of Tropical Medicine and Hygiene* 93(2): 380–383.

Coelho, Katia E. F. A., et al. 2018, Nov. "Congenital Zika syndrome phenotype in a child born in Brazil in December 2011." *Clinical Case Reports* 6(11): 2053–2056.

Dick, G. W.A. 1952, Sept. "Zika virus (I). Isolations and serological specificity." *Transactions of the Royal Society of Tropical Medicine and Hygiene* 46(5): 509–520.

Donofrio, Peter D. 2017, Oct. "Guillain-Barré syndrome." *Continuum (Minneapolis, Minn.)* 23(5, Peripheral Nerve and Motor Neuron Disorders): 1295–1309.

Duffy, Mark R., et al. 2009. "Zika virus outbreak on Yap Island, Federated States of Micronesia." *New England Journal of Medicine* 360: 2536–2543.

Fernandez-Garcia, Maria Dolores, Michela Mazzon, Michael Jacobs, and Ali Amara. 2009, Apr 23. "Pathogenesis of flavivirus infections: Using and abusing the host cell." *Cell Host and Microbe* 5(4): 318–328.

Foy, Brian D., et al. 2011, May. "Probable non-vector-borne transmission of Zika virus, Colorado, USA." *Emerging Infectious Diseases* 17(5): 880–882.

Gourinat, Ann Claire, et al. 2015, Jan. "Detection of Zika virus in urine." *Emerging Infectious Diseases* 21(1): 84–86.

Hafer-Macko, C. E., et al. 1996, May. "Immune attack on the schwann cell surface in acute inflammatory demyelinating polyneuropathy." *Annals of Neurology* 39(5): 625–635.

Kaur, Guneet, Pallavi Pant, Reshma Bhagat, and Pankaj Seth. 2023. "Zika virus E protein alters blood-brain barrier by modulating brain microvascular endothelial cell and astrocyte functions." *bioRxiv*: 2023.02.09.527854. http://biorxiv.org/content/early/2023/02/10/2023.02.09.527854.abstract.

Kuno, G., and Chang, G. J. J. 2007. "Full-length sequencing and genomic characterization of Bagaza, Kedougou, and Zika viruses." *Archives of Virology* 152(4): 687–696.

van der Linden, Vanessa, et al. 2016, Dec 2. "Description of 13 infants born during October 2015–January 2016 with congenital Zika virus infection without microcephaly at birth—Brazil." *MMWR. Morbidity and Mortality Weekly Report* 65(47): 1343–1348.

MacNamara, F. N. 1954, Mar. "Zika virus: A report on three cases of human infection during an epidemic of jaundice in Nigeria." *Transactions of the Royal Society of Tropical Medicine and Hygiene* 48(2): 139–145.

Marchette, N. J., Garcia, R., and A. Rudnick. 1969, May. "Isolation of Zika virus from aedes aegypti mosquitoes in Malaysia." *American Journal of Tropical Medicine and Hygiene* 18(3): 411–415.

Mulkey, Sarah B., et al. 2020, Mar. "Neurodevelopmental abnormalities in children with in utero Zika virus exposure without congenital Zika syndrome." *JAMA Pediatrics* 1;174(3): 269–276.

Musso, Didier, et al. 2015, Jul. "Detection of Zika virus in saliva." *Journal of Clinical Virology* 68: 53–55.

Nithiyanantham, Saiee F., and Alaa Badawi. 2019, Oct. "Maternal infection with Zika virus and prevalence of congenital disorders in infants: Systematic review and meta-analysis." *Canadian Journal of Public Health* 110(5): 638–648.

Olson, J. G., Ksiazek, T. G., Suhandiman, G., and Triwibowo, V. 1981. "Zika virus, a cause of fever in Central Java, Indonesia." *Transactions of the Royal Society of Tropical Medicine and Hygiene* 75(3): 389–393.

Pielnaa, Paul, et al. 2020, Apr. "Zika virus—Spread, epidemiology, genome, transmission cycle, clinical manifestation, associated challenges, vaccine and antiviral drug development." *Virology* 543: 34–42.

Pierson, Theodore C., and Margaret Kielian. 2013, Feb. "Flaviviruses: Braking the entering." *Current Opinion in Virology* 3(1): 3–12.

Prisant, Nadia, et al. 2016, Sept. "Zika virus in the female genital tract." *Lancet Infectious Diseases* 16(9): 1000–1001.

Sarno, Manoel, et al. 2016, Feb 25. "Zika virus infection and stillbirths: A case of hydrops fetalis, hydranencephaly and fetal demise." *PLoS Neglected Tropical Diseases* 10(2): e0004517.

Satterfield-Nash Ashley, Kotzky Kim, Allen Jacob, Bertolli Jeanne Bertolli, Moore Cynthia A., Pereira Isabela Ornelas, Pessoa André, Melo Flavio, Santelli Ana Carolina Faria e Silva, Boyle Coleen, A., Peacock Georgina. 2017. "Health and development at age 19–24 months of 19 children who were born with microcephaly and laboratory evidence of congenital Zika virus infection during the 2015 Zika virus outbreak—Brazil, 2017." *MMWR Morb Mortal Weekly Report* 66(49): 1347–1351.

Schuler-Faccini, Lavínia, et al. 2022, Mar 8. "Neurodevelopment in children exposed to Zika in utero: Clinical and molecular aspects." *Frontiers in Genetics* 13: 758715.

Shoenfeld, Yehuda, Jacob George, and James B. Peter. 1996, Apr. "Guillain-Barré as an autoimmune disease." *International Archives of Allergy and Immunology* 109(4): 318–326.

Wielanek, A. C., et al. 2007, Nov 27. "Guillain-Barré syndrome complicating a chikungunya virus infection." *Neurology* 69(22): 2105–2107.

World Health Organization. 2017. WHO WHO | *Zika Situation Report*.

Chapter 6

Neuropathological Outcomes in Human Influenza and Respiratory Syncytial Virus Infections

Sabyasachi Dash

6.1 INTRODUCTION

Influenza viruses belong to the *Orthomyxoviridae* family and are classified as either type A, B, C, or the recently identified type D (1). Influenza A viruses (IAVs) and type B viruses (IBVs) contain eight negative-sense, single-stranded viral RNA (vRNA) gene segments, as represented in Figure 6.1, which encode transcripts for ten essential viral proteins, as well as accessory proteins that are several strain-dependent (2,3). IAVs are further classified into 18 subtypes based on the combination of haemagglutinin (HA) and 11 subtypes based on neuraminidase (NA) glycoproteins on their surfaces. In comparison, influenza type C and D viruses only possess seven vRNA gene segments, as the hemagglutinin–esterase fusion protein coding vRNA replaces the HA and NA vRNAs. In this chapter we will discuss largely the IAV types, as they are the primary causative agents for flu pandemics where most of the neurological outcomes have been recorded. There is a multitude of subtypes or variants in the flu virus, however, only three have consistently affected the human population that has resulted in pandemics. These three strain variants include H1N1 (1918 and 2009 pandemics), H2N2 (1957 pandemic), and H3N2 (1968 pandemic) (4).

6.2 BINDING AND ENTRY OF INFLUENZA VIRUS

The binding and entry of flu virus to host cells is a multistep process (Figure 6.2). The binding of the virus to host cells is an HA-mediated binding to the receptor that triggers endocytosis of the virion. The endocytosis of the virus particle can occur either via the engagement of clathrin vesicles, involving dynamin and the adaptor protein (e.g., Epsin-1), or by macropinocytosis. Once the virus enters the cell, it is then trafficked to the endosome, where the low pH activates the M2 ion channel and causes a large conformational change in HA that exposes the fusion peptide. This opening of the M2 ion channel acidifies the internal space of the viral particle due to which the packaged viral ribonucleoproteins get released from M1. This step is important for the transfer of the viral proteins to the host cytoplasm

DOI: 10.1201/9781003285823-6

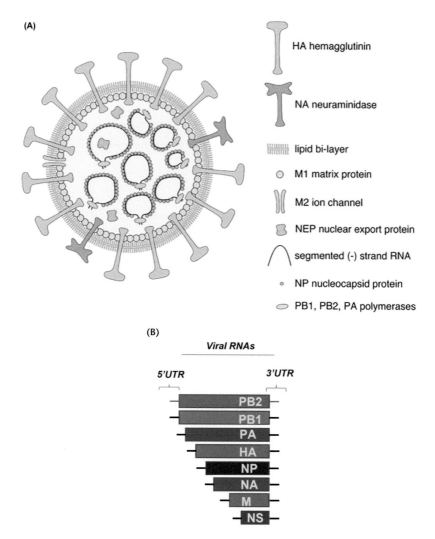

Figure 6.1 Structure of influenza virus. (A) Schematic representation of the human influenza A virus. (B) Schematic of the eight viral RNA (vRNA) gene segments that constitute the influenza A and B genomes. The 5′ and 3′ untranslated regions, which contain the viral promoters, are represented with a line, and the box corresponds to the coding region within each vRNA.

Source: Adapted from (5).

post-membrane fusion. The HA protein facilitates the fusion of the viral-endosomal membranes in a sequence of steps wherein the host proteases ultimately cleave the HA protein into two subunits HA1 and HA2. This cleavage is critical for the activation of the viral protein that enables the exposure of the fusion peptide on the N-terminus of the HA2 protein when

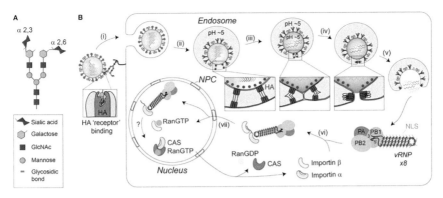

Figure 6.2 Receptor-mediated cell entry of influenza A virus. (A) Diagram of a bi-antennary N-linked glycan. The terminal sialic acid residues are displayed with an α-2,3 linkage, as well as an α-2,6 linkage, to illustrate the "linear" and "bent" presentations. (B) Illustration of IAV cell entry. (i) IAVs initiate cell entry by using the HA receptor-binding domain (located in the HA1 region) to associate with sialylated glycoconjugates on a host "receptor." Binding to the "receptor" triggers endocytosis. (ii) The virus then traffics to the endosome where the lower pH facilitates a conformational change in HA, exposing the fusion peptide (located in the HA2 region) for insertion into the endosomal membrane. (iii) The HA pre-hairpin conformation begins to collapse, forming a six-helix bundle that promotes hemifusion of the viral envelope with the endosomal membrane. At some point, the M2 channel opens to release the viral ribonucleoproteins (vRNPs) from M1 by acidifying the viral interior. (iv) HA further collapses into a trimer of hairpins to promote the formation of the fusion pore, which (v) releases the vRNPs into the cytosol. (vi) The exposed nuclear localization signals (NLS) on the vRNPs are recognized by the adaptor protein importin-α, leading to the recruitment of importin-β that (vii) facilitates the transport through the nuclear pore complex (NPC) and into the nucleus.

Source: Reprinted with permission from (6).

the pH changes in the endosomal vesicles. When this fusion peptide exposure takes place, it gets inserted into the endosomal membrane, while the C-terminal transmembrane domain remains anchored to the HA2 subunit in the viral membrane, creating a pre-hairpin conformation.

The HA2 trimers then fold back to create a hairpin that begins to position the two membranes near each other. The hairpin bundles then collapse into helix-like conformation, which allows the two membranes to come closer together, facilitating the subsequent fusion of the two membranes as depicted in Figure 6.2.

6.3 INFLUENZA PATHOGENESIS

The influenza virus has gone through genomic evolution over a period through gradual accumulation of mutations and genetic reassortment (Figure 6.3). As a result, this virus presents a significant healthcare burden in the human population every year, with approximately 15% of the global population becoming

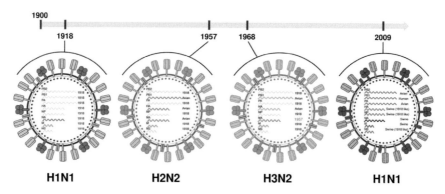

Figure 6.3 Evolution of influenza virus. There have been four influenza pandemics since the turn of the 20th century, occurring in 1918 (H1N1), 1957 (H2N2), 1968 (H3N2), and 2009 (H1N1). This timeline shows the temporal and genetic reassortment relationships among each of the pandemic influenza subtypes.

Source: Adapted from (7).

infected by this virus annually. Because of this natural genetic evolutionary process, the virus genome acquires mutations that helps it escape the host immune system and successfully establish infection in the human population. Once the virus enters the host cell, it counteracts hosts immune responses via its virulence factors. One such factor is the viral NS1 protein, which targets specific pathways of the host's innate response (8). Unlike any other viral infection, influenza pathogenicity triggers tissue injury and illness that is a consequence of the viral replication. As a result, several lines of immune signaling and host-induced responses are activated that can be observed through cytokine release and systemic inflammation. Research findings in animal models demonstrates that influenza infection causes pulmonary inflammation due to compromised lung epithelial cells, alveolar pneumocytes, and alveolar macrophages in the course of infection. The virus establishes infection by breaking down the alveolar barrier function of the pneumocytes. This causes alveolar flooding, influx of neutrophils and monocytes, as a result causing clogging of small airways with desquamated cells. In the murine model of influenza (H2N2 strain) pathogenesis, nitric oxide, and oxygen radicals have been shown to play a significant role in exacerbation of pulmonary injury (9). However, when it comes to the impact of this virus on the brain, the direct contribution of viral replication and infectivity cycle remains unclear.

6.4 NEUROLOGICAL OUTCOMES IN INFLUENZA PATHOGENESIS

Seasonal influenza–associated neurologic complications are generally categorized into groups based on the diagnosis in patients that may represent overlapping pathogenesis and severity of outcomes. These groups include febrile seizures, encephalopathy, encephalitis, Reye's-like syndrome, acute disseminated encephalomyelitis, and acute necrotizing encephalitis (ANE). Historically, 1918–1919 is considered one of the deadliest global influenza pandemics. Interestingly, this period was more than a decade prior to the discovery of the influenza virus that was first identified in the 1930s (10). Though the majority of deaths were mainly attributed to respiratory malfunctioning, only a spectrum of neurological illness in these flu patients was associated with this pandemic. Sporadic cases of brainstem encephalitis were reported in a cohort of patients from Vienna (11,12). Several influenza-positive patients presented with brainstem encephalitis with indications of lethargy, insomnia, sleep reversal, headache, vomiting, vertigo, dry mouth, hiccups, dysuria, and tremors; these symptoms lasted anywhere from a few days to weeks. Some patients also experienced paralysis or weakness of the eye muscles. Patients also developed progressive lethargy prior to entering a state of deep unconsciousness. In some patients this lethargic phenotype lasted approximately 2 weeks and gradually diminished naturally; in cases when it was severe, it progressed to a coma state (12). A subset of lethargic patients were observed to exhibit ocular abnormalities that ranged from ophthalmoplegia to pupillary malfunctioning. Facial palsy, hemiparesis, and cerebellar ataxia were also recorded in such conditions. The death rate was higher in such patients; however, in survivors the incidents of severe neurologic sequelae were uncommon. For example, in a study 112 encephalitis brain autopsies were clinically reviewed (13). In this study the median age was found to be 29 years and the median duration of the illness that patients had experienced was a maximum of 10 days. Results from the neuropathological examination revealed inflammation in the brainstem ranging from the midbrain to the medulla region. There were signs of lymphocytic perivascular cuffing, foci of parenchymal lymphocytes, foci of glial nodules, and vascular congestion in the brain regions. This encephalopathy phenotype observed in the patients has been reported in association with other influenza viral strains (H1, H2, and H3), including influenza B strains (14–17). In a study conducted on 842 patients hospitalized with influenza from 2000 to 2004, 10% of the patient population was observed to develop neurological symptoms (18). It is often discussed that the risk of encephalopathy is higher in children with pre-existing neurological problems. Infected children tend to develop focal neurologic signs such as hemiparesis, aphasia, muscle spasms, and backward arching of head, neck, and spine (19). Given the spectrum of neuropathologies that are observed in influenza-infected patients, radiologists have divided the types of MRI images into several categories: normal, diffuse involvement with edema of the cerebral cortex, symmetric involvement of the thalami, basal ganglia and brainstem (ANE), and a postinfectious or acute disseminated encephalomyelitis category with multiple abnormal foci, mainly in the white matter (20,21).

6.5 CLINICAL CASE REPORTS FOR INFLUENZA-ASSOCIATED BRAIN PATHOLOGY

Influenza-associated neuropathological outcomes exhibit a spectrum of indications. Some of these indications are more well recognized than others. The clinical manifestations can range from milder symptoms to fully severe with a chronic and aggressive phenotype culminating in mortality. Due to non-uniformity in the clinical presentation of influenza infections as well as lack of detection of the virus in the cerebrospinal fluid, accurate diagnosis becomes challenging. Moreover, there is a lack of diagnostic criteria in these individuals. Given the importance of this subject, in this section we have reviewed a handful of clinical case studies on the neurological pathologies recorded in the patient population (from pediatric to adulthood) as represented in Figures 6.4–6.7 with the aim to gain a better understanding of the neurological presentations.

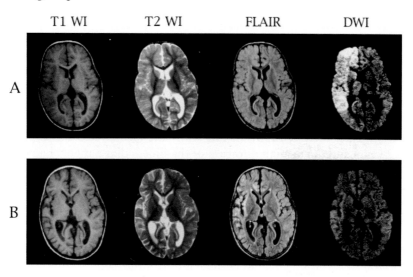

Figure 6.4 Magnetic resonance imaging (MRI) case report of the brain of a 20-month-old girl with fever, cough, and rhinorrhea. Results of brain computerized tomography scanning were normal. Intensive supportive therapy was performed after admission. Four days post-hospitalization, she had the second episode of hemiconvulsions on the left side, and her consciousness was further disturbed. (A) Acute and (B) convalescent phases of influenza-related encephalopathy or encephalitis. (A) Diffusion-weighted imaging (DWI) delineates the right hemispheric lesions better than conventional MRI (T1-weighted, T2-weighted, and fluid-attenuated inversion recovery [FLAIR] imaging) on the 8th hospital day. (B) Follow-up MRI on the 26th day shows significant improvement of the lesions. A FLAIR image in the convalescent phase reveals more increased signal intensity in the cortex of the right hemisphere than that in the acute phase.

Source: Figure reprinted with permission from (22).

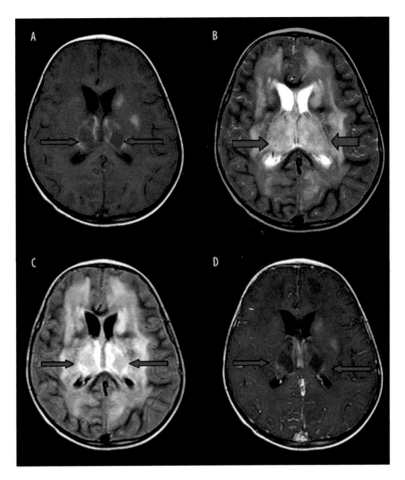

Figure 6.5 Case reports of influenza-associated neurological complications in children. *Top panel*: A 3-year-old boy with acute necrotizing encephalitis, with fever for 2 days and convulsion once. The thalamus was swollen. Symmetrical and multifocal involvement was observed in bilateral thalamus and paraventricular ([A] T1WI; [B] T2WI; [C] T2WI-FLAIR), and no enhancement was observed in all lesions ([D] T1WI contrast enhancement). Axial T1WI (A) showed the tricolor pattern in thalamic lesions. *Bottom panel*: An 11-month-old girl with influenza-associated encephalopathy, with fever for 2 days, no convulsion. Multiple gyrus swelling and abnormal signals were observed in bilateral cerebral hemispheres, with low signal on T1WI (A) and high signal on T2WI (B) and T2WI-FLAIR (C), more obviously on the left side. Diffuse enhancement was observed on T1WI contrast enhancement (D) in lesions of the gyrus and pia meninges.

Source: Figure reprinted with permission from (23).

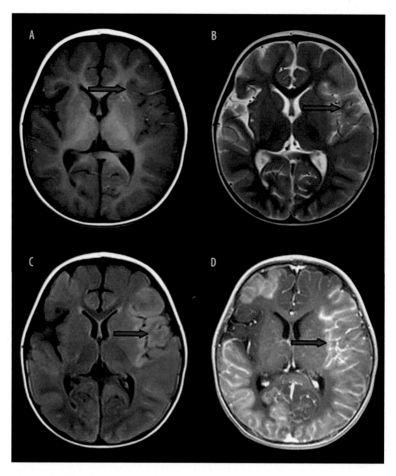

Figure 6.5 (Continued)

Most cases are pediatric in nature, recorded in children under age 6 years with the neurologic symptoms beginning within 1 to 3 days of the onset of respiratory signs (24,25). Adults can also develop ANE, and often the signs of lethargy with multiple seizures occur in children (24). Recent studies of serum and cytokine levels from patients with influenza who developed encephalopathy have suggested that elevated cytokine levels may be in part responsible for the neuroinflammatory phenotype. Studies of patients with encephalopathy have reported elevated serum and/or cerebrospinal fluid (CSF) elevated levels of cytokines such as the interleukins IL-1α, IL-6, IL-8, IL-10, IL-15, and TNF-α, and soluble TNF receptor (26,27). In a few autopsies, influenza viral antigen has been detected in brain tissue, in Purkinje cells and some pontine nuclei, and within capillary vessel walls of the basal ganglia (28–32).

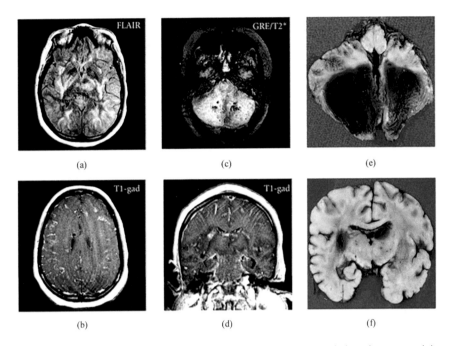

Figure 6.6 Fatal case of HINI-associated acute necrotizing encephalopathy in an adult. (a) FLAIR MRI sequence demonstrates widespread abnormal signal in the sub-cortical white matter, internal capsule, putamen, globus pallidus, and thalamus bilaterally. (b) T1 post-gadolinium axial MRI sequence demonstrates diffuse lep-tomeningeal enhancement over the convexities bilaterally and foci of parenchymal enhancement within the cortex bilaterally. (c) Gradient echo (GRE/T2*) MRI sequence demonstrates hemorrhage in the deep cerebellar white matter bilaterally (mottled black signal, representing magnetic susceptibility). (d) T1 post-gadolinium coronal MRI sequence demonstrates basal leptomeningeal enhancement surrounding the brainstem. (e) Postmortem axial section through the brainstem and cerebellum demonstrates bilateral hemorrhage and necrosis of the cerebellar hemispheres. (f) Postmortem coronal section demonstrates widespread but patchy areas of hemorrhage and necrosis in the deep gray nuclei bilaterally and the cerebral hemispheres bilaterally.

Source: Adapted and reprinted with permission from (33).

In 2009, a new H1N1 pandemic was the causative agent of high mortality rates and exhibited increased reports of neurological complications. According to this, a retrospective study of the clinical files of 55 patients infected with H1N1 detected a 50% of visible neurological symptoms (34). In this cohort, the most frequent neurological sign reported was headache (in 35% of the patients) as well as severe neurological complication (in 9% of the patients). Another study performed in Malaysia collected clinical data from pediatric hospitals during the 2009 pandemic. The study reported

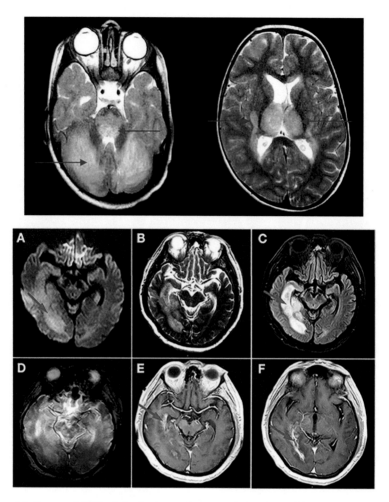

Figure 6.7 *Top panel*: MRI of the brain on the second day showed hyperintense changes in the thalami bilaterally, brainstem, cerebellum, and subcortical cortex. Axial magnetic resonance imaging brain scan in patient showing hyperintense lesion in thalami, cerebellum, and pons (*arrowed*). *Bottom panel*: Brain MRI. (A) Axial DWI and (B) axial T2WI showed a high signal, and the size of the edematous area showed a marked increase at the right temporal occipital lobe (*arrow*). (C) FLAIR image showed a swollen, edematous temporal occipital lobe (*arrow*). (D) Axial GRE sequence at the same level with the high signal of T2WI. (E, F) A contrast-enhanced image demonstrated an abnormal enhancement of the right temporal occipital lobe and bilateral choroid (*arrow*). DWI, diffusion-weighted imaging; FLAIR, fluid-attenuated inversion-recovery; GRE, gradient echo image; T2WI, T2-weighted image. Magnetic resonance imaging brain scan in patient showing hyperintense lesion in thalami, cerebellum, and pons (*arrowed*).

Sources: (Top Panel) Adapted and reprinted with permission from (35). (Bottom Panel) Adapted and reprinted with permission from (36).

that 8.3% of children under 5 years old presented neurological manifestation, among which more than 60% presented febrile seizures. Importantly,
approximately 15% of children exhibited influenza-associated encephalitis
and ANE, respectively. Brain imaging analysis revealed that a few patients
exhibited alterations such as cerebral edema and ANE, along with a handful
of cases where the neurological sequelae was permanent (37).

In another study with a cohort of 74 hospitalized children, seizure-related
neurologic complications (mainly seizures) were observed in 19% of the
children. Only one patient from this study was diagnosed with transverse
myelitis and presented permanent sequelae (38). In cases of fatality attributed to the H1N1 pandemic infection, clinical findings showed that the cause
of death was an intracerebral thrombosis and hemorrhage with presence of
the virus in the brain but not in lungs or CSF (39). The H3N2 and H1N1
seasonal flu have also been associated with neurological manifestations. In
a study with 21 patients with neurological alterations showed that the primary clinical sign was encephalitis, and about 50% of the patients have
sequelae (15). The presence of neuroradiological diagnosis suggests that the
phenotype observed in patients may be a cause of direct viral invasion into
the central nervous system (CNS) (40,41). Besides the neurological signs
described above, a study in Japan reveals that patients with neurological
manifestations also had mild impairment of consciousness, typically delirium or hallucinations and abnormal behavior, among others, which belongs
to the neuropsychiatric disorders (42,43). In Japan, influenza is the most
identified pathogen associated with acute encephalopathy, whereas in the
United States and Australia, approximately 6%–20% of children hospitalized with influenza infection have neurological manifestations (44–49).

6.6 HUMAN RESPIRATORY SYNCYTIAL VIRUS AND ITS NEUROLOGICAL OUTCOMES

The human respiratory syncytial virus (hRSV) is an enveloped, negative-
sense, single-stranded RNA virus that belongs to the *Mononegavirales* order
and has recently been assigned to the *Pneumoviridae* family and the *Ortho-
pneumovirus* genus. RSV infection can lead to serious respiratory problems in vulnerable populations, such as children younger than 1 year of
age and immunocompromised older adults. The 15.2 kb RSV genome is
a non-segmented negative-sense RNA that encodes for 11 viral proteins,
namely nonstructural proteins NS1 and NS2, nucleoprotein (N), phospho-
protein (P), matrix protein (M), small hydrophobic protein (SH), attachment glycoprotein (G), fusion protein (F), M2–1, M2–2, and large protein
(L) (Figure 6.8). The RSV envelope contains three surface transmembrane
glycoproteins, specifically G, F, and SH. RSV primarily targets airway epithelial cells wherein the binding and viral entry into host cells is mediated by the
G and F proteins. The G protein is responsible for viral attachment to host

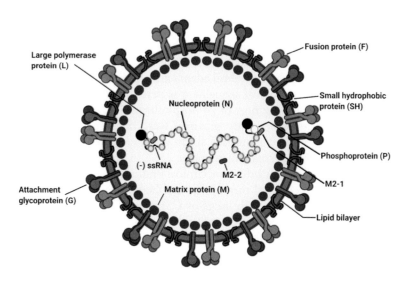

Figure 6.8 The structure of respiratory syncytial virus (RSV). The RSV genome is 15.2 kb of non-segmented, negative-sense RNA encoding 11 viral proteins. The viral envelope of RSV contains three transmembrane glycoproteins: attachment gly-coprotein (G), fusion protein (F), and small hydrophobic protein (SH). Matrix proteins (M) are present on the inner side of the viral envelope. Viral RNA is tightly protected by nucleoproteins (N) and the large polymerase protein (L), phosphoproteins (P), and M2–1 protein that mediates viral RNA transcription. M2–2 protein regulates viral RNA synthesis.

Source: Adapted and reprinted with permission from (50).

cells and immune modulation post-establishment of a successful infection. RSV entry is mediated by the F protein, which undergoes a conformational change and fuses the viral envelope with the host cell membrane to facilitate its entry into the host cell (50).

RSV is primarily responsible for bronchiolitis and pneumonia (51,52). However, there have been reports of extrapulmonary manifestations in HRSV infection that include myocarditis and encephalopathy (53–55). In a study performed in children with hRSV infection, a febrile status was detected that manifested in neurological damage. Viral antibodies against the human RSV have been detected in the CSF of patients who have suffered symptoms of CNS infection such as seizures, convulsions, and neck stiffness. Importantly, this report suggests that RSV infection was associated with neurological abnor-malities such as encephalitis. In a case report of an RSV-positive 3-year-old child, clinical signs of ataxia were observed (56). In this case report the authors concluded that the child manifested a meningoencephalitis phenotype with cerebellitis because of the viral infection. In a retrospective study published in the year 2001, clinical data from 487 RSV-positive patients with bronchiolitis were evaluated for neurological signs (57). The results of this analysis showed that 1.8% of the children exhibited visible clinical signs of encephalopathy (seizure phenotype). Another retrospective investigation evaluated 236 patients of whom 121 were RSV-positive and 115 were RSV-negative. In the RSV-positive cohort, about 6.6% presented seizures and approximately 20% of patients exhibited sleep apnea; however, this number was similar for the one reported in the RSV-negative cohort.

The first detection of RSV viral RNA in CSF was from a 4-month-old boy hospitalized with pneumonia and febrile convulsion (58). In this study, the authors were able to identify that the RSV strain found belonged to the serogroup B. In the CSF of an 11-month-old boy that exhibited neurological abnormalities, an increase of IL-6 in the CSF but not in serum was recorded. This finding suggests a tissue region specific effect implicating that CNS cell types (i.e., astrocytes and microglia) can be a potential source of these cytokines. Similar findings on elevated IL-6 levels were found in three cases of infants younger than 2 years, in which it was possible to detect RSV RNA (serogroup A) in the CSF (59). These findings led the authors to hypothesize a possible route of direct invasion of the CNS by the RSV. RSV infection is associated largely with encephalopathies that are classified in four groups: (1) metabolic error; (2) cytokine storm; (3) excitotoxicity; and (4) hypoxic (60). The encephalopathy caused by metabolic error is an abnormality of the brain function that can be reversed and involves an alteration of metabolites. In the second type of encephalitis, a high increase of several cytokines at systemic levels is observed which also affects other organs and tissues. In some of these patients exhibiting the cytokine storm, excitotoxic encephalopathy was found, which is characterized by febrile convulsion status epilepticus. Some RSV patients also manifested encephalopathies associated with hypoxia. Importantly, RSV RNA was found in CSF in at least 50% of the patients, and the levels of IL-6 were increased only in the patients who exhibited excitotoxic or cytokine storm encephalopathy type. Moreover, the study identified that in all the patients the levels of nitric oxide (NO) were significantly increased irrespective of the encephalopathy. These results are consistent with the previous report of these authors where they described the finding of RSV RNA in patients and elevated NO levels when compared to influenza-associated encephalopathies. It must be acknowledged that the frequency of neurological complications associated with RSV infection is low; however, the outcomes in the CNS remain consistent across patients (61).

6.6.1 Case Report for RSV Infection in a Pediatric Patient

As discussed earlier in this chapter, cerebral damage caused by RSV and perhaps concomitant cardiac involvement may result in sudden cardiac arrest in infants with bronchiolitis. Shown in Figure 6.9 is the case report of an

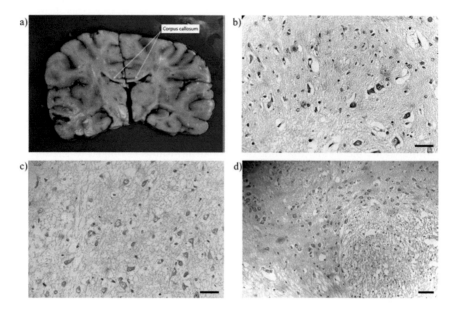

Figure 6.9 (a) Macroscopic and microscopic brain damage. Extensive tissue injury was observed in the brain, characterized by white matter degradation in the left hemisphere, basal ganglia, hippocampus, bulbar region, and pons with edema. (b) A detailed microscopic analysis showed hypertrophic astrocytes, acutely damaged glia, and focal necrosis at the bulbar level. (c, d) Basal ganglia were affected by neuronal necrosis with perineuronal halo and white matter strongly damaged.

Source: Figure reprinted with permission from (52).

infant from Italy admitted with RSV-related severe bronchiolitis and a consequential rapid neurological involvement, resulting in a fatal outcome (52).

6.7 CONCLUSION

It is obvious that respiratory viruses such as influenza and syncytial virus are driving agents of pneumonia and pulmonary malfunctions. The clinical manifestations are not restricted to any age group, however, they exhibit severe outcomes in children and the elderly due to low immunity. It is remarkable to note that these viruses also lead to extra-pulmonary outcomes in the form of neurological dysfunction and sequelae. Febrile seizures, loss of consciousness, convulsions, encephalitis, and even ataxia-like phenotypes and ocular dysfunction are among the several outcomes that are described in this chapter. These case reports from various infected patients across age groups spotlights the urgency to acknowledge the CNS impact and to highlight the need to better understand the molecular mechanisms that influence the neuroinvasive properties of these respiratory viruses.

REFERENCES

1. Hause BM, Collin EA, Liu R, Huang B, Sheng Z, Lu W, et al. Characterization of a novel influenza virus in cattle and swine: Proposal for a new genus in the orthomyxoviridae family (Palese P, editor). *MBio* [Internet] 2014;5(2). Available from: https://journals.asm.org/doi/10.1128/mBio.00031-14

2. Palese P, Schulman JL. Mapping of the influenza virus genome: Identification of the hemagglutinin and the neuraminidase genes. *Proceedings of the National Academy of Sciences* [Internet] 1976;73(6); 2142–2146. Available from: https://pnas.org/doi/full/10.1073/pnas.73.6.2142

3. McGeoch D, Fellner P, Newton C. Influenza virus genome consists of eight distinct RNA species. *Proceedings of the National Academy of Sciences* [Internet] 1976;73(9); 3045–3049. Available from: https://pnas.org/doi/full/10.1073/pnas.73.9.3045

4. Morens DM, Taubenberger JK, Fauci AS. The persistent legacy of the 1918 influenza virus. *New England Journal of Medicine* [Internet] 2009;361(3); 225–259. Available from: www.nejm.org/doi/abs/10.1056/NEJMp0904819

5. Kabiljo J, Laengle J, Bergmann M. From threat to cure: Understanding of virus-induced cell death leads to highly immunogenic oncolytic influenza viruses. *Cell Death Discovery* [Internet] 2020;6; 48. Available from: www.ncbi.nlm.nih.gov/pubmed/32542113

6. Dou D, Revol R, Östbye H, Wang H, Daniels R. Influenza a virus cell entry, replication, virion assembly and movement. *Frontiers in Immunology* [Internet] 2018;9. Available from: www.frontiersin.org/article/10.3389/fimmu.2018.01581/full

7. Harrington WN, Kackos CM, Webby RJ. The evolution and future of influenza pandemic preparedness. *Experimental & Molecular Medicine* [Internet] 2021;53(5); 737–749. Available from: www.nature.com/articles/s12276-021-00603-0

8. Hale BG, Albrecht RA, García-Sastre A. Innate immune evasion strategies of influenza viruses. *Future Microbiology* [Internet] 2010;5(1); 23–41. Available from: www.futuremedicine.com/doi/10.2217/fmb.09.108

9. Akaike T, Noguchi Y, Ijiri S, Setoguchi K, Suga M, Zheng YM, et al. Pathogenesis of influenza virus-induced pneumonia: Involvement of both nitric oxide and oxygen radicals. *Proceedings of the National Academy of Sciences of the United States of America* [Internet] 1996;93(6); 2448–2453. Available from: https://pnas.org/doi/full/10.1073/pnas.93.6.2448

10. Shope RE. Swine influenza. *Journal of Experimental Medicine* [Internet] 1931;54(3); 349–359. Available from: https://rupress.org/jem/article/54/3/349/10190/SWINE-INFLUENZA-I-EXPERIMENTAL-TRANSMISSION-AND

11. Foley PB. Encephalitis lethargica and the influenza virus. II. The influenza pandemic of 1918/19 and encephalitis lethargica: Epidemiology and symptoms. *Journal of Neural Transmission* [Internet] 2009;116(10); 1295–1308. Available from: http://link.springer.com/10.1007/s00702-009-0295-9

12. Reid AH, McCall S, Henry JM, Taubenberger JK. Experimenting on the past: The enigma of von economo's encephalitis lethargica. *Journal of Neuropathology & Experimental Neurology* [Internet] 2001;60(7); 663–670. Available from: https://academic.oup.com/jnen/article-lookup/doi/10.1093/jnen/60.7.663

13. Anderson LL, Vilensky JA, Duvoisin RC. Review: Neuropathology of acute phase encephalitis lethargica: A review of cases from the epidemic period. *Neuropathology and Applied Neurobiology* [Internet] 2009;35(5); 462–472. Available from: https://onlinelibrary.wiley.com/doi/10.1111/j.1365-2990.2009.01024.x

14. Paisley JW. Type A2 influenza viral infections in children. *Archives of Pediatrics and Adolescent Medicine* [Internet] 1978;132(1); 34. Available from: http://archpedi.jamanetwork.com/article.aspx?doi=10.1001/archpedi.1978.02120260036007

15. Steininger C, Popow-Kraupp T, Laferl H, Seiser A, Gödl I, Djamshidian S, et al. Acute encephalopathy associated with influenza a virus infection. *Clinical Infectious Diseases* [Internet] 2003;36(5); 567–574. Available from: https://academic.oup.com/cid/article/36/5/567/451994

16. Lin CH, Huang YC, Chiu CH, Huang CG, Tsao KC, Lin TY. Neurologic manifestations in children with influenza b virus infection. *Pediatric Infectious Disease Journal* [Internet] 2006;25(11); 1081–1083. Available from: https://journals.lww.com/00006454-200611000-00021

17. Ekstrand JJ, Herbener A, Rawlings J, Turney B, Ampofo K, Korgenski EK, et al. Heightened neurologic complications in children with pandemic H1N1 influenza. *Annals of Neurology* [Internet] 2010;68(5); 762–766. Available from: https://onlinelibrary.wiley.com/doi/10.1002/ana.22184

18. Newland JG, Laurich VM, Rosenquist AW, Heydon K, Licht DJ, Keren R, et al. Neurologic complications in children hospitalized with influenza: Characteristics, incidence, and risk factors.

Journal of Pediatrics [Internet] 2007;150(3); 306–310. Available from: https://linkinghub.elsevier. com/retrieve/pii/S0022347606011310

19. Amin R, Ford-Jones E, Richardson SE, MacGregor D, Tellier R, Heurter H, et al. Acute childhood encephalitis and encephalopathy associated with influenza. *Pediatric Infectious Disease Journal* [Internet] 2008;27(5); 390–395. Available from: https://journals.lww.com/00006454-200805000-00003

20. Kimura S, Ohtuki N, Nezu A, Tanaka M, Takeshita S. Clinical and radiological variability of influenza-related encephalopathy or encephalitis. *Pediatrics International* [Internet] 1998;40(3); 264–270. Available from: https://onlinelibrary.wiley.com/doi/10.1111/j.1442-200X.1998.tb01925.x

21. Studahl M. Influenza virus and CNS manifestations. *Journal of Clinical Virology* [Internet] 2003;28(3); 225–232. Available from: https://linkinghub.elsevier.com/retrieve/pii/S1386653203001197

22. Tokunaga Y, Kira R, Takemoto M, Gondo K, Ishioka H, Mihara F, et al. Diagnostic usefulness of diffusion-weighted magnetic resonance imaging in influenza-associated acute encephalopathy or encephalitis. *Brain and Development* [Internet] 2000;22(7); 451–453. Available from: https://linkinghub.elsevier.com/retrieve/pii/S0387760400001790

23. Song Y, Li S, Xiao W, Shen J, Ma W, Wang Q, et al. Influenza-associated encephalopathy and acute necrotizing encephalopathy in children: A retrospective single-center study. *Medical Science Monitor* [Internet] 2020; 26. Available from: www.medscimonit.com/abstract/index/idArt/928374

24. Morishima T, Togashi T, Yokota S, Okuno Y, Miyazaki C, Tashiro M, et al. Encephalitis and encephalopathy associated with an influenza epidemic in Japan. *Clinical Infectious Diseases* [Internet] 2002;35(5); 512–517. Available from: https://academic.oup.com/cid/article-lookup/doi/10.1086/341407

25. Togashi T, Matsuzono Y, Narita M, Morishima T. Influenza-associated acute encephalopathy in Japanese children in 1994–2002. *Virus Research* [Internet] 2004;103(1–2); 75–78. Available from: https://linkinghub.elsevier.com/retrieve/pii/S0168170204001157

26. Ichiyama T, Isumi H, Ozawa H, Matsubara T, Morishima T, Furukawa S. Cerebrospinal fluid and serum levels of cytokines and soluble tumor necrosis factor receptor in influenza virus-associated encephalopathy. *Infectious Diseases* [Internet] 2003;35(1); 59–61. Available from: www.tandfonline.com/doi/full/10.1080/0036554021000026986

27. To KKW, Hung IFN, Li IWS, Lee K, Koo C, Yan W, et al. Delayed clearance of viral load and marked cytokine activation in severe cases of pandemic H1N1 2009 influenza virus infection. *Clinical Infectious Diseases* [Internet] 2010;50(6); 850–859. Available from: https://academic.oup.com/cid/article-lookup/doi/10.1086/650581

28. Jang H, Boltz D, Sturm-Ramirez K, Shepherd KR, Jiao Y, Webster R, et al. Highly pathogenic H5N1 influenza virus can enter the central nervous system and induce neuroinflammation and neurodegeneration. *Proceedings of the National Academy of Sciences* [Internet] 2009;106(33); 14063–14068. Available from: https://pnas.org/doi/full/10.1073/pnas.0900096106

29. Dourmashkin RR, Dunn G, Castano V, McCall SA. Evidence for an enterovirus as the cause of encephalitis lethargica. *BMC Infectious Diseases* [Internet] 2012;12(1); 136. Available from: https://bmcinfectdis.biomedcentral.com/articles/10.1186/1471-2334-12-136

30. Fraňková V, Jirásek A, Tůmová B. Type A influenza: Postmortem virus isolations from different organs in human lethal cases. *Archives of Virology* [Internet] 1977;53(3); 265–268. Available from: http://link.springer.com/10.1007/BF01314671

31. Ishigami A, Kubo S, Ikematsu K, Kitamura O, Tokunaga I, Gotohda T, et al. An adult autopsy case of acute encephalopathy associated with influenza A virus. *Legal Medicine* [Internet] 2004;6(4); 252–255. Available from: https://linkinghub.elsevier.com/retrieve/pii/S1344622304000276

32. Takahashi M, Yamada T, Nakashita Y, Saikusa H, Deguchi M, Kida H, et al. Influenza virus-induced encephalopathy: Clinicopathologic study of an autopsied case. *Pediatrics International* [Internet] 2000;42(2); 204–214. Available from: http://doi.wiley.com/10.1046/j.1442-200x.2000.01203.x

33. Lee YJ, Smith DS, Rao VA, Siegel RD, Kosek J, Glaser CA, et al. Fatal H1N1-related acute necrotizing encephalopathy in an adult. *Case Reports Crit Care* [Internet] 2011;2011; 1–4. Available from: www.hindawi.com/journals/cricc/2011/562516/

34. Asadi-Pooya AA, Yaghoubi E, Nikseresht A, Moghadami M, Honarvar B. The neurological manifestations of H1N1 influenza infection; diagnostic challenges and recommendations. *Iranian Journal of Medical Sciences* [Internet] 2011;36(1); 36–39. Available from: www.ncbi.nlm.nih.gov/pubmed/23365476

35. Incecik F, Ozlem Hergüner M, Altunbasak S, Yıldızdas D, Antmen B, Özgür Ö, et al. Fatal encephalitis associated with novel influenza A (H1N1) virus infection in a child. *Neurological Sciences* [Internet] 2012;33(3); 677–679. Available from: http://link.springer.com/10.1007/s10072-011-0839-2

36. Dou Y, Li Y. Influenza A H3N2-associated meningoencephalitis in an older adult with viral RNA in cerebrospinal fluid: Case report. *Frontiers of Neurology* [Internet] 2022; 13. Available from: www.frontiersin.org/articles/10.3389/fneur.2022.874078/full

37. Muhammad Ismail HI, Teh CM, Lee YL. Neurologic manifestations and complications of pandemic influenza A H1N1 in Malaysian children: What have we learnt from the ordeal? *Brain and Development* [Internet]. 2015;37(1); 120–129. Available from: https://linkinghub.elsevier.com/retrieve/pii/S0387760414000850

38. Landau YE, Grisaru-Soen G, Reif S, Fattal-Valevski A. Pediatric neurologic complications associated with influenza A H1N1. *Pediatric Neurology* [Internet] 2011;44(1); 47–51. Available from: https://linkinghub.elsevier.com/retrieve/pii/S0887899410003851

39. Simon M, Hernu R, Cour M, Casalegno J-S, Lina B, Argaud L. Fatal Influenza A(H1N1)pdm09 encephalopathy in immunocompetent man. *Emerging Infectious Diseases* [Internet] 2013;19(6); 1005–1007. Available from: http://wwwnc.cdc.gov/eid/article/19/6/13-0062_article.htm

40. Paiva TM, Theotonio G, Paulino RS, Benega MA, Silva DBB, Borborema SET, et al. Influenza virus A(H3N2) strain isolated from cerebrospinal fluid from a patient presenting myelopathy post infectious. *Journal of Clinical Virology* [Internet] 2013;58(1); 283–285. Available from: https://linkinghub.elsevier.com/retrieve/pii/S1386653213002011

41. Paksu MS, Aslan K, Kendirli T, Akyildiz BN, Yener N, Yildizdas RD, et al. Neuroinfluenza: Evaluation of seasonal influenza associated severe neurological complications in children (a multicenter study). *Child's Nervous System* [Internet] 2018;34(2); 335–347. Available from: https://link.springer.com/10.1007/s00381-017-3554-3

42. Mizuguchi M. Influenza encephalopathy and related neuropsychiatric syndromes. *Influenza and Other Respiratory Viruses* [Internet] 2013;7; 67–71. Available from: https://onlinelibrary.wiley.com/doi/10.1111/irv.12177

43. Manjunatha N, Math S, Kulkarni G, Chaturvedi S. The neuropsychiatric aspects of influenza/swine flu: A selective review. *Indian Journal of Psychiatry* [Internet] 2011;20(2); 83. Available from: https://journals.lww.com/10.4103/0972-6748.102479

44. Okuno H, Yahata Y, Tanaka-Taya K, Arai S, Satoh H, Morino S, et al. Characteristics and outcomes of influenza-associated encephalopathy cases among children and adults in Japan, 2010–2015. *Clinical Infectious Diseases* [Internet] 2018;66(12); 1831–1837. Available from: https://academic.oup.com/cid/article/66/12/1831/4775244

45. Sugaya N. Influenza-associated encephalopathy in Japan. *Seminars in Pediatric Infectious Diseases* [Internet] 2002;13(2); 79–84. Available from: https://linkinghub.elsevier.com/retrieve/pii/S1045187002500510

46. Goenka A, Michael BD, Ledger E, Hart IJ, Absoud M, Chow G, et al. Neurological manifestations of influenza infection in children and adults: Results of a national British surveillance study. *Clinical Infectious Diseases* [Internet] 2014;58(6); 775–784. Available from: https://academic.oup.com/cid/article-lookup/doi/10.1093/cid/cit922

47. Jantarabenjakul W, Paprad T, Paprad T, Anugulruengkitt S, Pancharoen C, Puthanakit T, et al. Neurological complications associated with influenza in hospitalized children. *Influenza and Other Respiratory Viruses* [Internet] 2023;17(1). Available from: https://onlinelibrary.wiley.com/doi/10.1111/irv.13075

48. Frankl S, Coffin SE, Harrison JB, Swami SK, McGuire JL. Influenza-associated neurologic complications in hospitalized children. *Journal of Pediatrics* [Internet] 2021;239; 24–31.e1. Available from: https://linkinghub.elsevier.com/retrieve/pii/S0022347621007319

49. Khandaker G, Zurynski Y, Buttery J, Marshall H, Richmond PC, Dale RC, et al. Neurologic complications of influenza A(H1N1)pdm09: Surveillance in 6 pediatric hospitals. *Neurology* [Internet] 2012;79(14); 1474–1481. Available from: www.neurology.org/lookup/doi/10.1212/WNL.0b013e31826d5ea7

50. Jung HE, Kim TH, Lee HK. Contribution of dendritic cells in protective immunity against respiratory syncytial virus infection. *Viruses* [Internet] 2020;12(1); 102. Available from: www.mdpi.com/1999-4915/12/1/102

51. Piedimonte G, Perez MK. Respiratory syncytial virus infection and bronchiolitis. *Pediatric Review* [Internet] 2014;35(12); 519–530. Available from: http://pedsinreview.aappublications.org/lookup/doi/10.1542/pir.35-12-519

52. Bottino P, Miglino R, Pastrone L, Barbui AM, Botta G, Zanotto E, et al. Clinical features of respiratory syncytial virus bronchiolitis in an infant: rapid and fatal brain involvement. *BMC Pediatrics* [Internet]. 2021;21(1); 556. Available from: https://bmcpediatr.biomedcentral.com/articles/10.1186/s12887-021-03045-9

53. Bohmwald K, Gálvez NMS, Ríos M, Kalergis AM. Neurologic alterations due to respiratory virus infections. *Frontiers in Cellular Neuroscience* [Internet]. 2018; 12. Available from: www.frontiersin. org/article/10.3389/fncel.2018.00386/full

54. Chiriboga-Salazar NR, Hong SJ. Respiratory syncytial virus and influenza infections: The brain is also susceptible. *Journal of Pediatrics* [Internet] 2021;239; 14–15. Available from: https://linkinghub. elsevier.com/retrieve/pii/S0022347621008222

55. Millichap JJ, Wainwright MS. Neurological complications of respiratory syncytial virus infection: Case series and review of literature. *Journal of Child Neurology* [Internet] 2009;24(12); 1499–1503. Available from: http://journals.sagepub.com/doi/10.1177/0883073808331362

56. Hirayama K, Sakazaki H, Murakami S, Yonezawa S, Fujimoto K, Seto T, et al. Sequential MRI, SPECT and PET in respiratory syncytial virus encephalitis. *Pediatric Radiology* [Internet] 1999;29(4); 282–286. Available from: https://link.springer.com/10.1007/s002470050589

57. Ng Y, Cox C, Atkins J, Butler IJ. Encephalopathy associated with respiratory syncytial virus bronchiolitis. *Journal of Child Neurology* [Internet] 2001;16(02); 105. Available from: http://journals. bcdecker.com/CrossRef/showText.aspx?path=JCN/volume 16%2C 2001/issue 02%2C February/ jcn_2001_6913/jcn_2001_6913.xml

58. Zlateva KT, Van Ranst M. Detection of subgroup b respiratory syncytial virus in the cerebrospinal fluid of a patient with respiratory syncytial virus pneumonia. *Pediatric Infectious Disease Journal* [Internet] 2004;23(11); 1065–1066. Available from: http://journals.lww.com/00006454-200411000-00022

59. Kawashima H, Ioi H, Ushio M, Yamanaka G, Matsumoto S, Nakayama T. Cerebrospinal fluid analysis in children with seizures from respiratory syncytial virus infection. *Infectious Diseases* [Internet] 2009;41(3); 228–231. Available from: www.tandfonline.com/doi/full/10.1080/00365540802669543

60. Morichi S, Kawashima H, Ioi H, Yamanaka G, Kashiwagi Y, Hoshika A, et al. Classification of acute encephalopathy in respiratory syncytial virus infection. *Journal of Infection and Chemotherapy* [Internet] 2011;17(6); 776–781. Available from: https://linkinghub.elsevier.com/retrieve/pii/ S1341321X1170410X

61. Miyamoto K, Fujisawa M, Tsuboi T, Hirao J, Sugita K, Arisaka O, et al. Systemic inflammatory response syndrome and prolonged hypoperfusion lesions in an infant with respiratory syncytial virus encephalopathy. *Journal of Infection and Chemotherapy* [Internet] 2013;19(5); 978–982. Available from: https://linkinghub.elsevier.com/retrieve/pii/S1341321X13700710

Chapter 7

The Vascular Route as a Possible Gateway for Central Nervous System Pathology in SARS-CoV-2 Infection

Sabyasachi Dash and Anne Khodarkovskaya

7.1 INTRODUCTION

The health pandemic associated with severe acute respiratory syndrome coronavirus-2 (SARS-CoV-2), commonly known as COVID-19, has gathered widespread attention and concern with more than 765 million positive cases and mortality counts of more than 6.9 million worldwide (1). The new SARS-CoV-2 shares significant structural and genetic similarities with the SARS-CoV initially reported in the early 2000s with signs of respiratory illness and pulmonary injury (2–4). However, the emerging COVID-19 pathologies are dramatically severe with a profoundly damaging impact on multiple organ systems. Coronaviruses (CoVs) are enveloped, positive-sense single-stranded (ss)RNA viruses featuring a spiked surface with unique replication mechanism. The group comprises the *Coronaviridae, Arteriviridae, Mesoniviridae*, and *Roniviridae* families of viruses with large non-segmented RNA genomes of ~30 kilobases (5). The host-pathogen interaction initiated with the viral spike (S) protein and its cognate host cell receptor is the most well-characterized mechanism for viral entry into target host cell (5).

Like other coronaviruses, symptoms of COVID-19 infection include common cold, dry cough, muscular pain, and high fever (6,7). Thereafter, signs of severe pneumonia and respiratory tract infections followed by acute respiratory distress and sepsis-induced shock with a demanding need for ventilator criteria are observed (6,8). It is also reported that COVID-19 infection can lead to neurological manifestations due to its neuroinvasive potential with a significant impact on the vascular endothelium (9,10). Particularly, viruses causing respiratory illnesses can cause serious damage to the structure and function of the vascular and circulation system leading to multiorgan injury and failure. Other than lungs as the prime targets of the viral infection, cardiovascular complications and acute kidney injury account for additional pathologies. Like SARS-CoV-2 outcomes, acute cardiac injury, shock, and arrhythmia were present with higher prevalence amongst COVID-19 patients, requiring intensive care. Such cardiovascular complications including myocarditis, acute myocardial infarction, and exacerbation of heart failure have been previously reported in influenza infection. In a

DOI: 10.1201/9781003285823-7

recent case report from New Orleans, Louisiana, conducted on lung and cardiac autopsy samples of COVID-19 patients, notable monocyte aggregates with pulmonary thrombosis and focal myocyte degeneration were reported. Furthermore, in COVID-19 cases, acute kidney injury marked by altered serum creatinine levels and diffused proximal tubule injury was observed. For instance, a molecular investigation of autopsy samples (26 patients) from China revealed the presence of viral particles in the tubular epithelium and podocytes. Distinct erythrocyte aggregates causing capillary lumen occlusion and fibrin deposits in the glomerular capillaries were associated with the injured endothelium. In another recent study from Germany, a broad range of viral tropism for SARS-CoV-2 was identified with viral particles detected in the kidneys, liver, heart, brain, and blood of patients. Even in patients without pre-existing kidney disease, acute kidney injury related to COVID-19 is detected. Similarly, liver disease and liver dysfunction are reported for comorbidity in COVID-19 patients. Critically ill COVID-19 patients seem to have higher rates of liver dysfunction irrespective of pre-existing liver diseases. Abnormal levels of liver aminotransferases, such as alanine and aspartate aminotransferases, are sought to be associated with a high rate of liver dysfunction in COVID-19 patients.

In addition to the gamut of pathological outcomes in multiple organs, the central nervous system (CNS) is also one of the secondary targets of viral pathogenesis wherein the viral particles and toxins can be trafficked using the circulation system because of the vascular damage. In general, the viral-induced CNS pathologies and associated behavioral alterations have been well studied for several viruses including human immunodeficiency virus (HIV), measles virus, Japanese encephalitis virus, Zika virus, and Ebola virus (11–14). An overview of the well-studied viruses causing human respiratory illness, including the novel SARS-CoV-2, and their associated pathologies is discussed in the prior chapters.

7.2 CEREBROVASCULAR COMPLICATIONS OF COVID-19

Clinical evidence suggests the association of neurological manifestations with the severity of SARS-CoV-2 (COVID-19) infection in patients (15–17). The neurological manifestations include ischemic stroke and cerebral hemorrhage, reduced consciousness, encephalopathy, cerebral edema, delirium, nerve pain, and hypertension (Figure 7.1). Additionally, multifocal lesions in the thalamus with signs of hemorrhage were observed in COVID-19 patients (18). SARS-CoV-2 viral particles have been also detected in the heart, lung, and small intestine of patients with pre-existing medical conditions such as renal transplant, coronary artery disease, hypertension, diabetes, and obesity. Infiltration of immune cells followed by the presence of apoptotic bodies were detected in the endothelial rich arterial vessels and capillaries of heart, lung, and the lumen of small intestine in these patients (19). New case studies

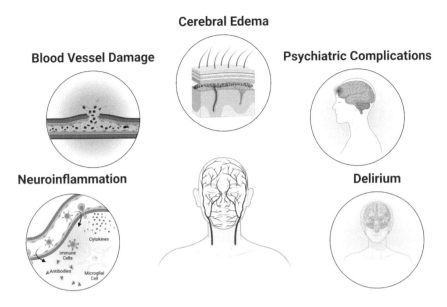

Cerebral Edema

Blood Vessel Damage

Psychiatric Complications

Neuroinflammation

Delirium

Figure 7.1 Common cerebrovascular complications in SARS-CoV-2 infection.

have reported on the incidence of large vessel ischemic stroke with elevated D-dimer levels, a marker for impaired vascular circulation in COVID-19 patients whose underlying mechanisms remain unknown (17,20). The initial symptoms in these patients included headache, cough, and lethargy, which with severity of infection progressed to signs of stroke such as sensory deficit, hemi-paralysis, and reduced consciousness. It was also reported that these patients were treated with clot retrieval medication such as tissue plasminogen activator (tPA) or aspirin.

To establish a successful infection, the SARS-CoV-2 virus largely depends on its S protein interaction with the host receptor angiotensin-converting enzyme 2 (ACE2) (Figure 7.2). ACE2 is a membrane-bound amino peptidase whose expression has been recorded in several integral tissue types including the lungs, arteries, heart, kidney, intestine, and various brain regions including the cerebral cortex, hypothalamus, brainstem, and the striatum (21–25). Summarizing these clinical observations confirms the presence of viral particles in the endothelial cells lining the blood vessels (19). Additionally, ACE2 mRNA and protein expression has been abundantly detected in the human lung and small intestine epithelia compared to arterial and venous endothelial cells (24,25). Thus, it is yet to be revealed whether the endothelial cells along the lining of blood vessels and of the several affected organs can be direct targets for SARS-CoV-2 infection or the virus uses additional mechanisms to directly infect endothelial cells. Given the sensitivity and technical advantage of single-cell RNA sequencing technology, multiple subtypes

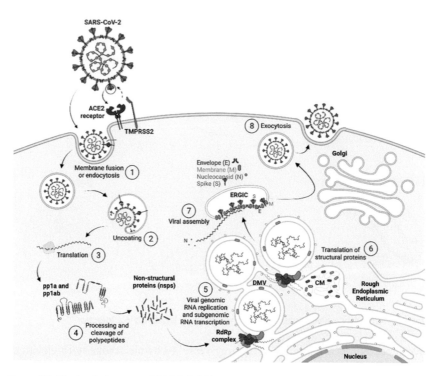

Figure 7.2 Mechanistic steps of SARS-CoV-2 entry in mammalian cells.

of a given cell population exist across tissues that prompt a speculation to identify the endothelial cell subtype specificity for SARS-CoV-2 infection. Another plausible hypothesis points to endothelial cells as indirect or secondary targets of the infected ACE2 expressing epithelial cells in the lungs.

The mechanistic steps of SARS-CoV-2 entry in mammalian cells are as follows (Figure 7.2):

1. Spike (S) proteins on the membrane of the SARS-CoV-2 virion mediate cellular entry by associating with angiotensin-converting enzyme 2 (ACE2) receptors on the cell surface, as well as other critical co-receptors. This results in viral entry through endosome or plasma membrane fusion. Host factor TMPRSS2, a cell surface serine protease, can cleave the spike protein and initiate fusion of the viral and plasma membranes.
2. Viral genomic RNA is released into the host cell via uncoating, revealing open reading frames ORF1a and ORF1b for translation.
3. Translation of the viral RNA results in polyproteins pp1a and pp1ab.
4. The polyproteins are cleaved in post-translational processing by proteases to form functional non-structural proteins (NSPs).

5. NSPs form the RNA replicase-transcriptase complex (RdRp). The endoplasmic reticulum (ER) forms a protective environment for RNA replication and transcription with double-membrane vesicles (DMVs) and convoluted membranes (CMs). RdRp works to replicate viral genomic RNA and transcribe genomic and subgenomic RNA.

6. Subgenomic mRNA is translated by ribosomes on the ER into structural and accessory proteins, including spike (S), envelope (E), and membrane (M) proteins.

7. The structural proteins are transported via the ER-to-Golgi intermediate compartment (ERGIC). Nucleocapsid (N) proteins in the cytoplasm are assembled from genomic RNA and associate with newly produced genomic RNA. Interaction between ERGIC proteins and N-encapsulated RNA results in viral assembly.

8. The assembled virion is transported through the Golgi apparatus and bud into small, smooth-walled vesicles. The virions are secreted at the cell surface through exocytosis.

Thereafter, viral infiltration into the bloodstream via the breach of endothelial barrier present along the lining of blood vessels causes long-lasting viremia that establishes significant neurotropic effects (26,27). The "Trojan horse" model of viral pathogenesis in the CNS is the most widely accepted model for virus-induced neuropathology. The viral particles traverse through the blood–brain barrier (BBB) via transcytosis and paracellular mechanisms, predominantly via the destabilization of endothelial tight junctions (27,28). The hijacking of these transport mechanisms by viral particles from the bloodstream to the brain parenchyma is exacerbated by the induced activity of the matrix metalloproteinase (MMP) enzymes leading to severe BBB dysfunction and a continuous infection cycle (27,28). Given this model, it can be hypothesized that COVID-19-associated vascular and neurological outcomes could be a result of alterations in the transcytotic and paracellular mechanisms followed by endothelial barrier dysfunction where the viral particles were detected in the tissue specific endothelial cells followed by large infiltration of immune cells and induced apoptotic marker expression.

7.3 VASCULAR AND INFLAMMATORY OUTCOMES IN VIRAL-INDUCED SIRS

A perturbed endothelial physiology is marked by increased vascular permeability, induction of pro-coagulant phenotype, and recruitment of monocytes and immune mediators in response to pathogenic or viral invasion. Thereafter, dissemination of the pathogen and immune factors via the microcirculation leads to sustained inflammation-hypoxia signaling, fibrin deposition, and blood vessel occlusion, a phenomenon known as the disseminated intravascular coagulation (DIC). DIC is commonly seen in sepsis, a condition manifested by abnormal blood flow and arteriovenous shunting leading to systemic inflammatory immune response (SIRS), which usually leads to life-threatening multiple organ dysfunction culminating in multiple organ failure (29,30). Respiratory viruses have a wide range of pathological outcomes given their broad tropism. Overall, the pathology of severe viral infections is like the development of SIRS. Hence it can be postulated that the pathological outcomes of SARS-CoV-2 often linked to multi-organ failure could be an outcome of the virus-induced SIRS. One such example could be the imbalance in the pro- versus anti-inflammatory cytokines is a one of the hallmarks of viral-induced immune dysfunction, is like SIRS pathology.

The sustained reduction in the peripheral lymphocyte population (aka lymphopenia) followed by a cytokine storm is associated with a very high risk of developing secondary infections (bacterial or viral) and may pose a critical factor associated with disease severity and mortality. However, the mechanisms underlying lymphopenia in viral pathologies remain unknown. In summary, studies have reported the presence of viral particles in blood, lymphoid tissues, and endothelial cells lining the blood vessels, suggesting the infectivity of SARS-CoV-2. Studies also indicate a significant percentage of abnormal intravascular coagulation marked by increased D-dimer and fibrin degradation in COVID-19 patients. Changes in the endothelial cell physiology by endotoxins, viral and bacterial pathogens cause intravascular coagulation causing severe endothelial injury and tissue damage across several organs leading to the detection and appearance of pathogens and immune elements in the circulation stream. Given this knowledge, the molecular mechanisms for abnormal coagulation and sepsis in SARS-CoV-2 infection deserves special attention. Being one of the first targets of pathogenic invasion in the circulation system, endothelium is a major target of sepsis and sepsis-induced events underlying SIRS pathology. Therefore, the dissemination of viral particles and immune factors from the lung to multiple organs can be postulated to occur via the endothelial activation lining the blood vessels along with the monocyte recruitment and platelet dysfunction in the circulation that causes sepsis and multiple organ failure.

7.4 CASE REPORTS OF CEREBROVASCULAR DYSFUNCTION AND NEUROLOGICAL IMPAIRMENT IN COVID-19 INFECTIONS

A 9-year-old girl was admitted to the pediatric intensive care unit (PICU) with a history of high-grade fever for 14 days, throbbing frontal headache, vomiting, and progressive weakness on the right side of her body for 5 days (Figure 7.3). This patient's condition deteriorated with worsening clinical outcomes such as bradycardia, and cardiac arrest on the third day of PICU admission. The patient was subjected to computed tomography (CT) angiography scan that revealed multifocal smooth stenosis of both intracranial

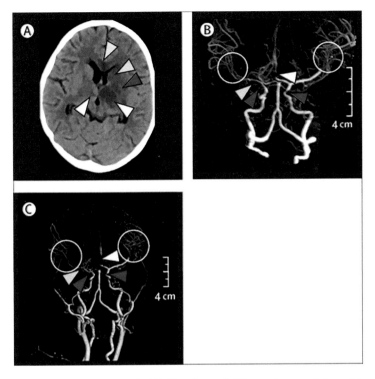

Figure 7.3 CT scan and angiograph. (A) Unenhanced CT scan image of the axial section showing hypodensities in corpus callosum (*green arrowhead*), left caudate (*yellow arrowhead*), putamen (*red arrowhead*), and bilateral thalami (*white arrowheads*), with mild compression over the left lateral ventricle. (B) Minimum intensity projection image of CT angiography, showing substantial stenosis of the anterior cerebral artery (*white arrowhead*), stenosis of bilateral supraclinoid internal carotid artery segments (*red arrowheads*), diffuse stenosis of M1 segment of the right middle cerebral artery (MCA; *yellow arrowhead*) and diffuse narrowing of the M2 and M3 segments of both MCA (*circles*). (C) Volume-rendered images of CT angiography showing multifocal narrowing involving both internal carotid arteries (*red arrowheads*), anterior cerebral artery (*white arrowhead*), and right MCA (yellow arrowhead). Note the diffuse narrowing of the segment M2 and M3 of both MCAs (*circled*); on the left side there is sudden tapering of the MCA at the M1–M2 junction. Scans taken on day 7 in pediatric intensive care unit.

Source: Reprinted with permission from (31).

internal carotid arteries, right middle cerebral artery, and anterior cerebral arteries as shown in Figure 7.3.

In another example from the many studies that have been reported, an international collaborative effort was launched to better understand the neurological manifestations in pediatric COVID patients. In this study, authors included a sample size of 38 patients from France, the United Kingdom, the

United States, India, Argentina, Peru, Brazil, and Saudi Arabia. Collectively, the pathologies seen in patients ranged from immune-mediated acute disseminated encephalomyelitis-like changes, myelitis, and neuronal alterations as well as intense lesions in the brains, as shown in Figures 7.4 and 7.5.

Another study reported a magnetic resonance imaging (MRI)-based hemorrhagic outcome in an adult infected with SARS-CoV-2 (Figure 7.6). Similarly, there are numerous reports highlighting the hemorrhagic and microvascular damage in the brain of COVID-infected patients as obtained from imaging

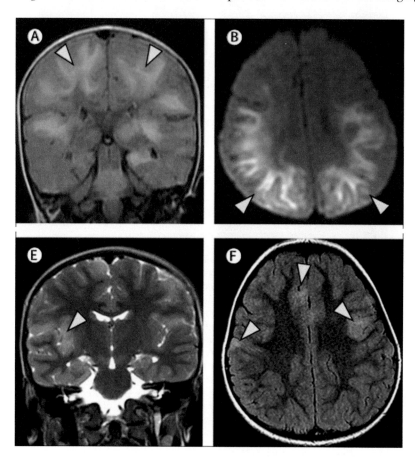

Figure 7.4 Acute disseminated encephalomyelitis (ADEM)-like brain changes. (A, B) A 1-year-old boy (case 2) with acute COVID-19 showed confluent areas of high signal in the subcortical white matter on coronal FLAIR imaging (A; *arrows*), and reduced diffusion on DWI trace (B; *arrows*). (E, F) In a 4-year-old boy (case 38) with an indeterminate timepoint of exposure to SARS-CoV-2, ADEM-like changes were seen on coronal T2-weighted images (E; *arrow*) and axial FLAIR images (F; *arrows*). This child was positive for antibodies to myelin oligodendrocyte glycoprotein. DWI, diffusion-weighted imaging. FLAIR, fluid-attenuated inversion recovery.

Source: Reprinted with permission from (32).

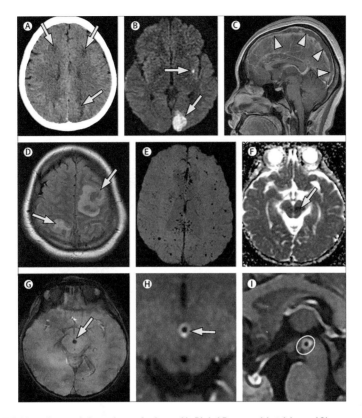

Figure 7.5 Vasculitis and thrombotic findings. (A, B) A 15-year-old girl (case 12) presented 27 weeks' pregnant with fever, seizures, and hypertension and COVID-19 pneumonia. Her CT at presentation (A) showed low-density areas in multiple locations (*arrows*). MRI 7 days later (B) showed small focal infarcts and a larger left occipital infarct (*arrows*) on diffusion trace imaging, findings compatible with unusually severe posterior reversible encephalopathy syndrome. (C, D) A 15-year-old girl (case 20) with subacute COVID-19 and no classical respiratory symptoms presented with fever, confusion, and headache. Complete occlusive thrombosis of the superior sagittal sinus was shown by the large filling defect in the post-contrast sagittal T1-weighted image (C; *arrowheads*), with resultant bilateral hemorrhagic venous infarcts on axial FLAIR images (D; *arrows*). (E) A 15-year-old girl (case 27) with multisystem inflammatory syndrome in children who also developed multiple microthrombi, as shown on SWI. The microthrombi were relatively clinically silent and showed partial resolution at 3 weeks with full clinical resolution of symptoms at 3 months after presentation. (F–I) A 2-year-old girl (case 32, indeterminate category) presented with fever and pharyngeal pain with an acute left midbrain infarction (*arrow*) shown on the apparent diffusion coefficient map (F; *arrow*). She had a thrombus in the feeding anterior perforator vessel on SWI (G; *arrow*) and marked associated vessel wall enhancement on postcontrast T1 arterial wall imaging (H, *arrow*; I, *circle*). FLAIR, fluid-attenuated inversion recovery. SWI, susceptibility-weighted imaging.

Source: Reprinted with permission from (32).

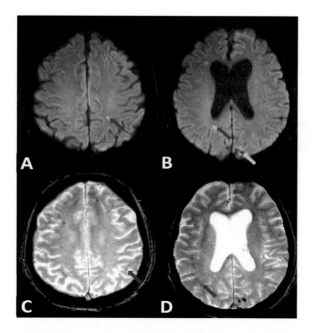

Figure 7.6 MRI images from a 61-year-old man with COVID-19. (A, B) Acute infarcts within the bilateral cerebral white matter (*blue arrows*) and left occipital hemorrhagic infarct (*green arrow*; images obtained using axial diffusion sequence); (C, D) innumerable microhemorrhages throughout the bilateral cerebral hemispheres (*red arrows*; images obtained using gradient-echo sequences).

Source: Reprinted with permission from (33).

analysis (Figures 7.7–7.10). These imaging findings are correlative to the molecular changes seen with neuroinflammatory and immune cell phenotypes that are associative with microvascular and neuronal dysfunction in the brain, as reported in the histological findings from the COVID-autopsy brain (Figure 7.10).

7.5 VASCULAR ENDOTHELIUM AS A PROPOSED THERAPEUTIC TARGET

Viral infection through the ACE2 receptor has been sought to damage the endothelial cells in the inner lining of the blood vessels causing amplification of proinflammatory cascade and perpetuating the infection (5,25,29,34,35). Endothelial dysfunction induced disassembly of cellular junctions causing barrier disruption and extravasation of leukocytes could be an explanation for the endothelial dysfunction and lymphopenia observed in COVID-19 patients (35,36). Endothelial dysfunction triggers a pro-coagulant state with increased proinflammation (i.e., cytokine storm) and oxidative stress that

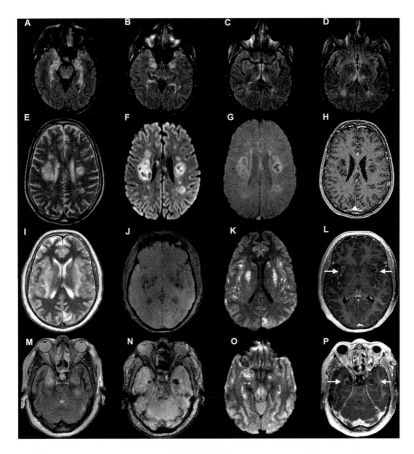

Figure 7.7 Imaging from patients with COVID-19 autoimmune and hemorrhagic encephalitis. Axial MRI from three individuals with para-/post-infectious central
syndromes. (A–D): Axial fluid-attenuated inversion recovery (FLAIR) images
show bilateral hyperintensity in the mesial temporal lobes (A, B), hypothalamus (C) temporal lobes and thalamus (D). (E–H): Axial T2-weighted (E),
diffusion-weighted imaging (DWI) (F), susceptibility-weighted imaging (SWI)
(G) and post-contrast T1-weighted (H) images show multifocal clusters of
lesions involving the deep white matter of both cerebral hemispheres, intral-
esional cyst-like areas of varied sizes, and some peripheral rims of restricted
diffusion (F), some hemorrhagic changes (G), and T1 hypointense "black holes"
without contrast enhancement (H). (I–P): Axial images at the level of the insula
and basal ganglia (I–L) and at the level of the temporal lobes and upper pons
(M–P). T2-weighted images (I and M), SWI images (J and N), DWI images (K, O)
and contrast-enhanced images (L, P). There are extensive confluent areas of T2
hyperintensity (I, M), with hemorrhagic change on SWI imaging (J, N), restricted
diffusion on DWI images (K, O) and peripheral contrast enhancement (arrows
in L and P) in the insular region, basal ganglia and left occipital lobe (I–L) as well
as in the medial temporal lobes and upper pons (M–P).

Source: Reprinted with permission from (37).

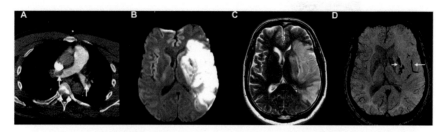

Figure 7.8 Imaging from patient 27, with cerebral infarction (A–D), and patient 41, with microhemorrhages (E–H). (A–D) Patient 27: CT pulmonary angiogram (A) demonstrated large emboli in the right and left pulmonary arteries (*arrows*). DWI (B), T2-weighted FSE (C) and, SWI (D) images show restricted diffusion (B) and T2 hyperintensity (C) in the left basal ganglia and cortical territory of left middle cerebral artery. The SWI image (D) shows hemorrhagic transformation in the basal ganglia (*short arrow*) and a long intravascular thrombus in a Sylvain branch of the left middle cerebral artery (*long arrow*).

Source: Reprinted with permission from (37).

causes apoptotic cell death, as observed in the patients. Thus, targeting the endothelial dysfunction factors and proinflammatory immune mediators could help overcome vascular outcomes. For instance, the sphingosine-1-phosphate receptor modulator Fingolimod (FTY720) licensed for multiple sclerosis is currently being repurposed and tested in Phase II for COVID-19 patients with acute respiratory distress (https://ClinicalTrials.gov Identifier NCT04280588). Several groups and research institutions worldwide are attempting to target the vascular endothelial cells with innovative strategies to combat the viral pathogenesis. In an interesting trial, Eli Lilly is attempting to target angiopoietin 2 (ANGPT2) using a monoclonal antibody approach in pneumonic patients with higher risk of progressing to acute respiratory distress syndrome (ARDS). ANGPT2 is a circulating ligand and vascular growth factor that antagonizes the endothelial cell specific tyrosine receptor kinase. Previous studies have established induced ANGPT2 levels in patients with ARDS, pneumonia, and chronic pulmonary injury (38–41). Additionally, bevacizumab, a monoclonal antibody targeting vascular endothelial growth factor, is also under investigation for COVID-19 patients with pulmonary edema and pneumonia (https://ClinicalTrials.gov Identifiers NCT04275414 and NCT04305106). Another endothelial specific protein sphingosine kinase 2 (SPHK2) has currently drawn attention since it has been shown to facilitate the replication of single-stranded RNA viruses and inhibition of SPHK2 activity using small molecule inhibitor Opaganib (aka Yeliva) ameliorated the levels of IL-6 and TNF-α in rodent model of influenza pathogenesis (42). Moreover, Menarini Silicon Biosystems is attempting to test its innovative liquid biopsy technology that involves the use of circulating endothelial cells to rescue the damaged lung vasculature in COVID-19 patients (30).

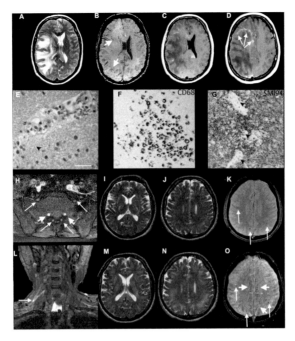

Figure 7.9 Axial MRI (A–D) and histopathology (E–G) from patient diagnosed with ADEM, and imaging (H–O) from patient, with combined CNS and PNS disease. (A–G) Axial T2-weighted (A), SWI (B), post-gadolinium (C and D) images show extensive confluent lesions involving the white matter of the right cerebral hemisphere, corpus callosum and corona radiata with mass effect, subfalcine herniation (A), clusters of prominent medullary veins (B, *short arrows*) and peripheral rim enhancement (D, *arrows*). (E) The white matter shows scattered small vessels with surrounding infiltrates of neutrophils and occasional foamy macrophages extending into the parenchyma (*arrow*). The endothelium is focally vacuolated but there is no evidence of vasculitis or fibrinoid vessel wall necrosis in any region. There were a few perivascular T cells in the white matter, but the cortex appears normal (not shown). (F) CD68 stain confirms foci of foamy macrophages in the white matter, mainly surrounding small vessels. There was no significant microgliosis in the cortex (not shown). (G) Myelin basic protein stain (SMI94) shows areas with focal myelin debris in macrophages around vessels in the white matter (*arrows*) in keeping with early myelin breakdown. There is no evidence of axonal damage on neurofilament stain (not shown). Scale bars: E = 45 μm; F and G = 70 μm. (H–O) Axial post-gadolinium fat-suppressed T1-weighted images (H) demonstrating pathologically enhancing extradural lumbosacral nerve roots (*arrows*). Note physiological enhancement of nerve root ganglia (*short arrows*). Coronal short tau inversion recovery (STIR) image (L) shows hyperintense signal abnormality of the upper trunk of the right brachial plexus (*arrow*). Initial axial T2 (I and J) and T2*-weighted images (K) show multifocal confluent T2 hyperintense lesions involving internal and external capsules, splenium of corpus callosum (I), and the juxtacortical and deep white matter (J), associated with microhemorrhages (K, *arrows*). Follow-up T2-weighted images (M and N) show marked progression of the confluent T2 hyperintense lesions, which involve a large proportion of the juxtacortical and deep white matter, corpus callosum and internal and external capsules. The follow-up SWI image (O) demonstrates not only the previously seen microhemorrhages (*arrows*) but also prominent medullary veins (*short arrows*).

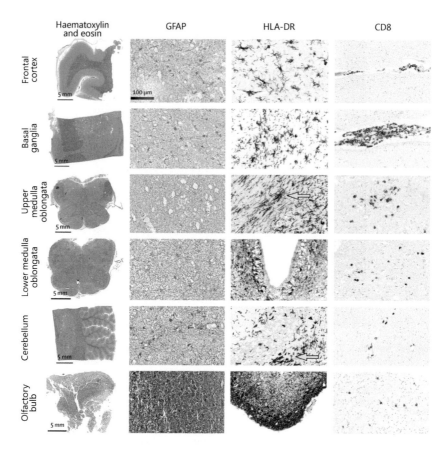

Figure 7.10 Molecular analysis of neuroinflammatory markers in COVID patient brain autopsy. An overview of each brain region with hematoxylin and eosin staining is shown in the first column. Immunohistochemical staining for the astrocytic marker GFAP showed variable degrees of reactive astrogliosis. Immunohisto-chemical staining for the microglia marker HLA-DR showed reactive activation of the microglia with occasional microglial nodules in the medulla oblongata and cerebellum (*green arrows*). Staining for the cytotoxic T lymphocyte marker CD8 (*brown*) revealed perivascular and parenchymal infiltration with CD8-positive cells. GFAP, glial fibrillary acidic protein.

Source: Reprinted with permission from (37).

7.6 CONCLUDING REMARKS

Even though a direct causal association between COVID-19 and stroke is not established, it is suggested that acute ischemic stroke reported in COVID-19 patients could be a result of severely induced prothrombotic state (17,43–45). The use of tPA or low-molecular-weight heparin for COVID-associated ischemic stroke treatment clearly indicates an imbalance in the body's pro-anticoagulant system which can be supported by the intravascular coagulation observed in non-survivors and critically ill patients (17,43–45). For instance, the activation of coagulation pathways by thrombin could augment proinflammatory response via activation of proteinase-activated receptors (46). Therefore, use of anticoagulants could be another attractive therapeutic approach to overcome the lung injury and vascular outcomes associated with COVID-19. Given such mounting evidence, the potential role of endothelial cells and mechanisms underlying endothelial dysfunction in COVID-19 pathology demands further investigation. A thorough investigation of the changes in endothelial function could also help identification of biomarkers unique to SARS-CoV-2 infection. Taken together, the ongoing situation opens a new window to better understand the role of vascular endothelium in SARS-CoV-2 pathology.

REFERENCES

1. World Health Organization/COVID-19. www.who.int/emergencies/diseases/novel-coronavirus-2019/situation-reports
2. Jaimes, J.A., André, N.M., Chappie, J.S., *et al*. Phylogenetic analysis and structural modeling of SARS-CoV-2 spike protein reveals an evolutionary distinct and proteolytically sensitive activation loop. *Journal of Molecular Biology* 2020, S0022–2836(20)30287–4. Advance online publication. https://doi.org/10.1016/j.jmb.2020.04.009
3. Zhang, T., Wu, Q., Zhang, Z. Probable pangolin origin of SARS-CoV-2 associated with the COVID-19 outbreak. *Current Biology: CB*. 2020, 30(8), 1578. https://doi.org/10.1016/j.cub.2020.03.063
4. Andersen, K.G., Rambaut, A., Lipkin, W.I., *et al*. The proximal origin of SARS-CoV-2. *Nature Medicine* 2020, 26, 450–452. https://doi.org/10.1038/s41591-020-0820-9
5. Fehr, A.R., Perlman, S. Coronaviruses: An overview of their replication and pathogenesis. *Methods in Molecular Biology (Clifton, NJ)* 2015, 1282, 1–23. https://doi.org/10.1007/978-1-4939-2438-7_1
6. Huang, C., et al. Clinical features of patients infected with 2019 novel coronavirus in Wuhan, China. *Lancet (London, England)* 2020, 395(10223), 497–506. https://doi.org/10.1016/S0140-6736(20)30183-5
7. Lim, Y.X., Ng, Y.L., Tam, J.P., *et al*. Human coronaviruses: A review of virus-host interactions. *Diseases (Basel, Switzerland)* 2016, 4(3), 26. https://doi.org/10.3390/diseases4030026
8. Cascella, M., Rajnik, M., Cuomo, A., *et al*. Features, evaluation and treatment Coronavirus (COVID-19) [Updated 2020 Apr 6]. In *StatPearls* [Internet]. Treasure Island (FL): StatPearls Publishing; 2020. Available from: www.ncbi.nlm.nih.gov/books/NBK554776/
9. Helms, J., Kremer, S., Merdji, H., *et al*. Neurologic features in severe SARS-CoV-2 infection. *New England Journal of Medicine* 2020, Jun 4, 382(23), 2268–2270. https://doi.org/10.1056/NEJMc2008597. Epub 2020 Apr 15. PMID: 32294339.
10. Li, Y.C., Bai, W.Z., Hashikawa, T. The neuroinvasive potential of SARS-CoV2 may play a role in the respiratory failure of COVID-19 patients. *Journal of Medical Virology* 2020, 92, 552–555. https://doi.org/10.1002/jmv.25728
11. Dash, S., Balasubramaniam, M., Villalta, F., *et al*. Impact of cocaine abuse on HIV pathogenesis. *Frontiers in Microbiology* 2015, 6, 1111. https://doi.org/10.3389/fmicb.2015.01111
12. Yin, Z., Wang, X., Li, L., *et al*. Neurological sequelae of hospitalized Japanese encephalitis cases in Gansu province, China. *American Journal of Tropical Medicine and Hygiene* 2015, 92(6), 1125–1129. https://doi.org/10.4269/ajtmh.14-0148
13. Howlett, P.J., Walder, A.R., Lisk, D.R., et al. Case series of severe neurologic sequelae of Ebola virus disease during epidemic, Sierra Leone. *Emerging Infectious Diseases* 2018, 24(8), 1412–1421. https://doi.org/10.3201/eid2408.171367 (Medline:30014839)

14. Blázquez, A.B., Saiz, J.C. Neurological manifestations of Zika virus infection. *World Journal of Virology* 2016, 5(4), 135–143. https://doi.org/10.5501/wjv.v5.i4.135

15. Mao, L., et al. Neurologic manifestations of hospitalized patients with coronavirus disease 2019 in Wuhan, China. *JAMA Neurology* 2020, 4, 2168–6157. https://doi.org/10.1001/jamaneurol.2020.1127

16. Pleasure, S.J., Green, A.J., Josephson, S.J. The spectrum of neurologic disease in the severe acute respiratory syndrome coronavirus 2 pandemic infection: Neurologists move to the frontlines. *JAMA Neurology* 2020, 4, 2168–6157. https://doi.org/10.1001/jamaneurol.2020.1065

17. Moore, H.B., Barrett, C.D., Moore, E.E., *et al.* Is there a role for tissue plasminogen activator as a novel treatment for refractory COVID-19 associated acute respiratory distress syndrome? *Journal of Trauma and Acute Care Surgery* 2020 (Volume Publish Ahead of Print—Issue). https://doi.org/10.1097/TA.0000000000002694

18. Neo, P., Gassan, S., Daniel, N., *et al.* COVID-19–Associated acute hemorrhagic necrotizing encephalopathy: CT and MRI features. *Radiology* 2020, Aug, 296(2), E119–E120. https://doi.org/10.1148/radiol.2020201187. Epub 2020 Mar 31. PMID: 32228363.

19. Varga, Z., *et al.* Endothelial cell infection and endotheliitis in COVID-19. *Lancet* 2020, May 2, 395(10234), 1417–1418. https://doi.org/10.1016/S0140-6736(20)30937-5. Epub 2020 Apr 21. PMID: 32325026.

20. Oxley, T.J., Mocco, J., Majidi, S., *et al.* Large-vessel stroke as a presenting feature of COVID-19 in the young. *New England Journal of Medicine.* 2020. https://doi.org/10.1056/NEJMc2009787

21. Wang, Q., Zhang, Y., Wu, L., *et al.* Structural and functional basis of SARS-CoV-2 entry by using human ACE2. *Cell.* 2020, S0092–8674(20)30338-X. https://doi.org/10.1016/j.cell.2020.03.045

22. Hoffmann, M., Kleine-Weber, H., Schroeder, S., *et al.* SARS-CoV-2 cell entry depends on ACE2 and TMPRSS2 and is blocked by a clinically proven protease inhibitor. *Cell.* 2020, 181(2), 271–280.e8. https://doi.org/10.1016/j.cell.2020.02.052

23. Li, M., Li, L., Zhang, Y. Expression of the SARS-CoV-2 cell receptor gene *ACE2* in a wide variety of human tissues. *Infectious Diseases of Poverty.* 2020, Apr 28, 9(1), 45. https://doi.org/10.1186/s40249-020-00662-x. PMID: 32345362. https://doi.org/10.1186/s40249-020-00662-x

24. Suryawanshi, H., Morozov, P., Muthukumar, T., *et al.* Thomas tuschl cell-type-specific expression of renin-angiotensin-system components in the human body and its relevance to SARS-CoV-2 infection. *bioRxiv.* https://doi.org/10.1101/2020.04.11.034603

25. Hamming, I., Timens, W., Bulthuis, M.L., *et al.* Tissue distribution of ACE2 protein, the functional receptor for SARS coronavirus. A first step in understanding SARS pathogenesis. *Journal of Pathology.* 2004, 203(2), 631–637. https://doi.org/10.1002/path.1570

26. Meyding-Lamadé, U., Craemer, E., Schnitzler, P. Emerging and re-emerging viruses affecting the nervous system. *Neurological Research and Practice.* 2019, 1, 20. https://doi.org/10.1186/s42466-019-0020-6

27. Miner, J.J., Diamond, M.S. Mechanisms of restriction of viral neuroinvasion at the blood-brain barrier. *Current Opinion in Immunology.* 2016, 38, 18–23. https://doi.org/10.1016/j.coi.2015.10.008

28. World Health Organization. Draft landscape of COVID-19 candidate vaccines as of 20th April 2020, www.who.int/blueprint/priority-diseases/key-action/novel-coronavirus-landscape-ncov.pdf

29. Tay, M. Z., et al. The trinity of COVID-19: Immunity, inflammation and intervention. *Nature Reviews Immunology* 2020, Jun, 20(6), 363–374. https://doi.org/10.1038/s41577-020-0311-8. Epub 2020 Apr 28. PMID: 32346093.

30. Tang, N., Li, D., Wang, X., *et al.* Abnormal coagulation parameters are associated with poor prognosis in patients with novel coronavirus pneumonia. *Journal of Thrombosis and Haemostasis: JTH.* 2020, 18(4), 844–847. https://doi.org/10.1111/jth.14768

31. Lindan, C.E., Mankad, K., Ram, D., *et al.* Neuroimaging manifestations in children with SARS-CoV-2 infection: A multinational, multicentre collaborative study. *Lancet. Child & Adolescent Health* 2021, 5(3), 167–177. https://doi.org/10.1016/S2352-4642(20)30362-X

32. Samkaria, A., Mandal, P.K. Brain imaging in COVID-19. *ACS Chemical Neuroscience* 2021, 12(16), 2953–2955. https://doi.org/10.1021/acschemneuro.1c00467

33. Paterson, R.W., Brown, R.L., Benjamin, L., *et al.* The emerging spectrum of COVID-19 neurology: Clinical, radiological and laboratory findings. *Brain: A Journal of Neurology* 2020, 143(10), 3104–3120. https://doi.org/10.1093/brain/awaa240

34. Li, H., *et al.* SARS-CoV-2 and viral sepsis: Observations and hypotheses. *Lancet (London, England)* 2020, S0140–S6736. https://doi.org/0.1016/S0140-6736(20)30920-X

35. Zhou, F., Yu, T., Du, R., *et al.* Clinical course and risk factors for mortality of adult inpatients with COVID-19 in Wuhan, China: A retrospective cohort study. *Lancet* 2020, 395(10229), 1054–1062.

36. Meyer, N.J., Calfee, C.S. Novel translational approaches to the search for precision therapies for acute respiratory distress syndrome. *Lancet. Respiratory Medicine*. 2017, 5(6), 512–523. https://doi.org/10.1016/S2213-2600(17)30187-X

37. Matschke, J., Lütgehetmann, M., Hagel, C., Sperhake, J.P., Schröder, A.S., Edler, C., Mushumba, H., Fitzek, A., Allweiss, L., Dandri, M., Dottermusch, M., Heinemann, A., Pfefferle, S., Schwabenland, M., Sumner Magruder, D., Bonn, S., Prinz, M., Gerloff, C., Püschel, K., Krasemann, S., Aepfelbacher, M., Glatzel, M. Neuropathology of patients with COVID-19 in Germany: A post-mortem case series. *Lancet. Neurology* 2020, 19(11), 919–929. https://doi.org/10.1016/S1474-4422(20)30308-2

38. Ando, M., Miyazaki, E., Abe, T., *et al*. Angiopoietin-2 expression in patients with an acute exacerbation of idiopathic interstitial pneumonias. *Respiratory Medicine*. 2016, 117, 27–32.

39. Su, L., Zhai, R., Sheu, C.C., *et al*. Genetic variants in the angiopoietin-2 gene are associated with increased risk of ARDS. *Intensive Care Medicine*. 2009, 35(6), 1024–1030. https://doi.org/10.1007/s00134-009-1413-8

40. Meyer, N.J., *et al*. ANGPT2 genetic variant is associated with trauma-associated acute lung injury and altered plasma angiopoietin-2 isoform ratio. *American Journal of Respiratory and Critical Care Medicine*. 2011, 183, 1344–53. https://doi.org/10.1164/rccm.201005-0701OC

41. Red Hill Biopharma Press Release. 27th April 2020. https://ir.redhillbio.com/news-releases/news-release-details/six-covid-19-patients-treated-redhills-opaganib-under

42. Menarini Silicon Biosystems Press Release 29th April 2020. www.menarini.com/Home/Menarini-News/News/News   details/ArticleId/2672/Menarini-Silicon-Biosystems-Explores-New-Application-for-Circulating-Endothelial-Cells-in-COVID-19.

43. José, R.J., Williams, A.E., Chambers RC Proteinase-activated receptors in fibroproliferative lung disease. *Thorax*. 2014, 69, 190–192

44. Wichmann, D., Sperhake, J., Lütgehetmann, M., *et al*. Autopsy findings and venous thromboembolism in patients with COVID-19: A prospective cohort study. *Annals of Internal Medicine* 2020 [Epub ahead of print 6 May 2020]. https://doi.org/10.7326/M20-2003

45. Tiwari, L., Shekhar, S., Bansal, A., *et al*. COVID-19 associated arterial ischaemic stroke and multisystem inflammatory syndrome in children: A case report. *Lancet Child Adolesc Health*. 2021, 5(1), 88–90. https://doi.org/10.1016/S2352-4642(20)30314-X

46. Burks, J.S., DeVald, B.L., Jankovsky, L.D., *et al*. Two coronaviruses isolated from central nervous system tissue of two multiple sclerosis patients. *Science*. 1980, 209, 933–934.10.1126/science.7403860

Chapter 8

Mechanisms for Viral Entry and Invasion in the Brain

Reshma Bhagat

8.1 INTRODUCTION

Viruses are infectious agents that rely on host cells to reproduce. Their primary goal is to invade a new host cell and introduce their genetic material into it. By doing so, they can take control of the cellular machinery and utilize it to replicate their own genetic material and produce more virus particles. In order to achieve this, the virus commonly exploits the host cell's endocytic mechanisms to enter the cell. Endocytosis is a process by which cells engulf external substances by forming a vesicle around them (Riezman et al. 1997). Viruses often take advantage of the host cell's natural endocytic mechanisms, which are involved in internalizing materials from the extracellular environment, to gain entry into the cell. They may bind to specific receptors on the cell surface, triggering the internalization process and allowing the virus to enter the cell. Subsequently, they utilize the host cell's cytoplasmic transport systems to migrate towards a replication site located within the cytosol (in the case of most RNA viruses) or nucleus (in the case of most DNA viruses) (Rampersad and Tennant 2018). The viral genome's protective coating is typically shed as the concluding step of this process. Many viruses have a protective coating or envelope surrounding their genetic material. This coating helps shield the virus from the host cell's defense mechanisms. However, during the infection process, the virus needs to shed or remove this protective coating to allow its genetic material to interact with the host cell's machinery and initiate replication. This step is crucial for the virus to continue its life cycle inside the infected cell.

The primary sites of viral entry encompass six key locations: the plasma membrane, early endosome, maturing endosome, late endosome, macropinosome, and endoplasmic reticulum (ER). Each of these locations offers distinct advantages and interactions with the host cell's molecular machinery. The plasma membrane serves as a pivotal entry point for viruses. By binding to specific receptors on the cell surface, viruses can initiate attachment and subsequent entry. Depending on the virus, this entry can occur through membrane fusion, whereby the viral envelope merges with the plasma membrane, or receptor-mediated endocytosis, whereby the virus is engulfed by the cell and enclosed in a membrane-bound vesicle.

Once inside the host cell, some viruses are transported to early endosomes. These compartments are involved in the internalization and sorting

DOI: 10.1201/9781003285823-8

of materials from the cell surface. Within the early endosome, the virus can exploit the acidic environment and other factors to undergo further internalization or uncoating. The acidic pH triggers conformational changes in viral proteins, leading to the release of the viral genome from its protective coating.

As endosomes progress through the endocytic pathway, they mature and undergo changes in composition and acidity. Maturing endosomes provide an environment where certain viruses can continue their internalization or uncoating processes. Late endosomes, which participate in the processing and sorting of cellular cargo, also offer a conducive environment for viral entry. These compartments can harbor viruses and provide access to the host cell, enabling viral replication and infection.

In addition to the endocytic pathway, viruses can exploit macropinocytosis as an entry pathway. Macropinocytosis involves the engulfment of fluid or large particles by the host cell. Certain viruses induce the formation of macropinosomes, which engulf the viral particles and transport them into the host cell's interior. This entry mechanism allows viruses to bypass specific receptors and gain access to the host cell in a non-selective manner.

Furthermore, some viruses possess the capability to directly target the ER, a network of membranes involved in protein synthesis and processing. By entering the ER, viruses can utilize their resources for replication, assembly, and interaction with cellular components. This strategy is employed by viruses with complex replication cycles or those requiring the ER for proper protein folding and maturation.

While the six locations represent the primary routes of viral entry, some viruses may also exploit other cellular compartments such as endolysosomes, amphisomes, and lysosomes. However, the evidence supporting viral entry through these alternative locations remains limited. Moreover, certain viruses exhibit the ability to utilize multiple pathways for entry. They achieve this by interacting with different receptors on the host cell surface, which allows them to exploit various entry routes and increase their chances of successful infection. Understanding the diversity of locations and pathways involved in viral entry into host cells provides valuable insights into the intricacies of viral infection.

The process of a virus entering a host cell is complex and can vary depending on the virus and host cell. However, it often involves the virus binding to a specific receptor on the host cell surface and being taken up through endocytic pathways. These pathways include clathrin-mediated, caveolin-mediated, clathrin- and caveolin-independent, and macropinocytosis. Once inside the host cell, the virus can be transported to various organelles, such as endosomes, lysosomes, and the Golgi apparatus, for further processing and replication. In some cases, the virus can also enter the host cell through fusion of its own membrane with the host cell membrane.

8.2 CLATHRIN-MEDIATED ENDOCYTOSIS

Clathrin-mediated endocytosis is a process by which cells internalize extra-cellular materials, such as nutrients, signaling molecules, and receptors, by forming clathrin-coated vesicles (CCVs). Clathrin is a protein that forms a lattice structure around the vesicle, which helps to deform the plasma membrane and bud off the vesicle into the cell's cytoplasm. The process of clathrin-mediated endocytosis involves several steps, including:

1. *Initiation*: The process is initiated by the binding of cargo molecules or receptors to specific receptors on the plasma membrane.
2. *Assembly of clathrin coat*: The adaptor proteins, such as AP2, recruit clathrin molecules to the plasma membrane, forming a clathrin coat around the cargo molecules and invaginating the membrane.
3. *Budding*: Once the clathrin coat is formed, the plasma membrane invaginates further and forms a vesicle containing the cargo molecule. Dynamin, a GTPase protein, mediates the scission of the vesicle from the plasma membrane.
4. *Uncoating*: The clathrin coat is then removed from the vesicle by a chaperone protein, Hsc70, allowing the vesicle to merge with endosomes and lysosomes. Clathrin-mediated endocytosis plays a critical role in a variety of cellular processes, including receptor-mediated endocytosis, regulation of cell surface signaling, and the uptake of nutrients.

Clathrin-mediated endocytosis can also be utilized by viruses to enter host cells (Figure 8.1). Many viruses, such as influenza virus, adenovirus, Zika virus, and HIV, use clathrin-mediated endocytosis as their primary entry pathway. The viral particles first attach to specific receptors on the cell surface, which triggers the recruitment of clathrin and adaptor proteins to the site of attachment. The clathrin then assembles around the viral particle, forming a coated pit, which invaginates the plasma membrane and forms a CCV containing the virus. Once inside the cell, the virus may then be delivered to different compartments, such as early endosomes, late endosomes, or lysosomes, depending on the virus type. In some cases, the low pH within these compartments triggers the release of the viral genome into the cytoplasm, where it can begin to replicate and produce new viral particles.

The viral surface protein of influenza virus, hemagglutinin interacts with sialic acid receptors on the host cell surface, leading to the internalization of the virus in CCVs. Vesicular stomatitis virus, a member of the *Rhabdoviridae* family, also enters host cells through clathrin-mediated endocytosis. The viral glycoprotein G interacts with specific receptors on the cell surface, triggering CCV formation and subsequent viral entry. The viral spike (S) protein of severe acute respiratory syndrome coronavirus (SARS-CoV), binds to ACE2 receptors on the host cell surface, leading to clathrin-dependent internalization. Middle East respiratory syndrome coronavirus (MERS-CoV), another coronavirus that causes severe respiratory illness, enters host cells through a similar mechanism

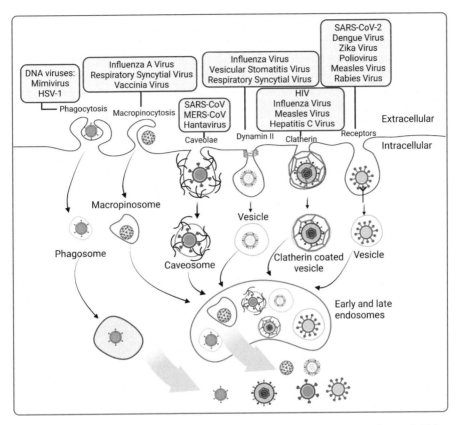

Figure 8.1 RNA viruses employ various cellular mechanisms for their entry into host cells. This includes clathrin-mediated endocytosis, receptor-mediated entry, caveolae-mediated entry, macropinocytosis, dynamin-mediated endocytosis, and even phagocytosis in certain cases. These mechanisms allow the viruses to bind to specific receptors on the cell surface, leading to internalization into the host cell. Clathrin-mediated endocytosis involves the binding of the virus to specific cell surface receptors, leading to the formation of clathrin-coated pits and subsequent internalization of the virus. Macropinocytosis is a process in which cells engulf large volumes of extracellular fluid and particles. Phagocytosis, typically associated with professional phagocytes, can also be utilized by some DNA viruses for their entry into host cells. Dynamin-mediated endocytosis involves the activity of the protein dynamin, which promotes the internalization of the virus into the cell. Some RNA viruses also utilize receptor-mediated entry, where they bind to specific cell surface receptors to enter the host cell.

involving clathrin-mediated endocytosis. The viral spike protein interacts with DPP4 receptors on the host cell surface, initiating CCV formation.

8.3 CAVEOLAE-MEDIATED ENDOCYTOSIS

Caveolae are distinct membrane domains that are enriched in cholesterol and sphingolipids, giving them unique lipid composition compared to the surrounding plasma membrane. These lipid-rich microdomains are involved in diverse cellular processes including endocytosis, lipid metabolism, and signaling. They act as signaling hubs by organizing and compartmentalizing signaling molecules, including G-proteins, receptor tyrosine kinases, and various downstream effectors. The clustering of signaling molecules within caveolae facilitates efficient signal transduction and can influence cellular processes such as cell growth, survival, and differentiation.

Caveolae are composed of caveolin proteins, with three major isoforms identified in mammalian cells: Cav-1, Cav-2, and Cav-3. These caveolins provide the structural framework for caveolae formation and are responsible for their characteristic morphology. Each isoform exhibits tissue-specific expression patterns and may have distinct roles in caveolae biology.

Cav-1 is a 178-amino acid protein that is highly hydrophobic and contains two hydrophobic domains near its N- and C-termini. Both termini are oriented to the cytoplasmic side and connected by a central hydrophobic domain that is inserted into, but does not span, the membrane bilayer in a hairpin configuration. Interestingly, a peptide corresponding to the last 20 residues of the hydrophilic N-domain, enriched in aromatic residues, can also bind to membranes independently. This region is known as the caveolin scaffolding domain and is a highly conserved region responsible for many functions associated with Cav-1. One of the pathways that has been described involves the internalization of viruses through caveolae, which are specialized lipid rafts enriched with the protein caveolin-1. Caveolae-mediated endocytosis is characterized by the formation of flask-shaped invaginations in the plasma membrane, which eventually pinch off to form caveolar vesicles. Once inside the cell, the viral particles can cluster receptor molecules such as certain integrins or glycosphingolipids within the lipid rafts, which increases their affinity for these domains and facilitates their entrapment.

Once the virus particles have clustered the receptor molecules within lipid rafts, they induce a cascade of tyrosine phosphorylation reactions. This phosphorylation cascade ultimately leads to rearrangements of the cortical actin cytoskeleton, which is important for the formation and function of caveolae.

The rearrangements of the cortical actin cytoskeleton allow the caveolae to pinch off from the plasma membrane and become internalized within the cell as endocytic vesicles. These vesicles then fuse with the endosomal membrane, allowing the virus particles to enter the cell. Initially, the virus is found in early endosomes and subsequently becomes associated with late endosomes resembling multivesicular bodies and possibly endolysosomes. Endosome maturation involves acidification of the compartments, which is required for simian vacuolating virus 40 (SV40) subsequent transport steps and the initiation of productive infection.

There are some notable exceptions of virus trafficking diverting from this pathway, despite being internalized through caveolae. For instance, cellular penetration of the EV1 virus, a positive-stranded RNA human pathogen, depends on caveolins, dynamin II, and signaling events but does not require actin filaments or microtubules. The virus uptake is much faster than that of SV40 and was followed by rapid co-localization with Cav-1. Beyond this step, the virus does not enter the Golgi complex, the ER, or the lysosomes. This observation raised the hypothesis that the virus particles remain sequestered in the Cav-1 positive endocytic vesicles until replication is initiated, further using them for cytoplasmic penetration and uncoating.

The precise molecular details characterizing the trafficking pathways of viruses are still being investigated, and understanding the factors involved in these unusual trafficking pathways is crucially important as other, yet uninvestigated viruses may well use them when accessing the host cell via caveolae. Moreover, there is accumulating evidence suggesting that several viruses take advantage of crosstalk between endocytosis routes, and that they can follow alternative trafficking pathways to reach their target organelles. The detailed understanding of these pathways will aid in the development of antiviral strategies and treatments.

8.4 NON–CLATHRIN DEPENDENT AND NON–CAVEOLAE DEPENDENT PATHWAYS OF VIRUS ENTRY

In addition to clathrin-mediated endocytosis and caveolin-mediated endocytosis, there are several other pathways that viruses can use to enter host cells. Some of these pathways do not require clathrin or caveolin coats, and are known as non-clathrin, non-caveolin endocytosis pathways. These pathways include macropinocytosis, flotillin, Arf6, cdc42-mediated endocytosis, and tyrosine kinase-dependent endocytosis.

Macropinocytosis is a process by which cells internalize extracellular fluids and large molecules through the formation of large vesicles known as macropinosomes. This process is initiated by the extension of actin-rich membrane protrusions called ruffles, which wrap around extracellular material and internalize it within the cell. The formation of macropinosomes is a non-specific process that occurs in response to a variety of stimuli, including growth factors, cytokines, and pathogens such as viruses. Macropinocytosis has been shown to be an important route of entry for several viruses. For example, vaccinia virus, the causative agent of smallpox, can activate a cellular signaling pathway that leads to the induction of macropinocytosis and subsequent viral entry. This pathway is initiated by the virus binding to cellular receptors, which triggers the activation of a host kinase called p21-activated kinase 1 (PAK1). PAK1 then phosphorylates and activates several downstream effectors, which leads to the formation of ruffles and the subsequent internalization of the virus within macropinosomes.

Once inside the macropinosome, the virus must overcome several barriers to reach the cytoplasm and establish infection. The macropinosome undergoes a maturation process that involves acidification of the lumen, fusion with sorting vesicles, and ultimately fusion with lysosomes, which can lead to degradation of the virus. However, some viruses have evolved strategies to escape this fate and instead utilize the macropinocytic pathway as a means of transport to the cytoplasm. For example, the adenovirus capsid can disrupt the macropinosome membrane and release its contents into the cytoplasm, where the viral genome can be delivered to the nucleus and initiate infection.

Flotillin-mediated endocytosis involves the lipid raft–associated protein flotillin and is dependent on dynamin-2 (DNM2) and cholesterol. Viruses such as respiratory syncytial virus and vaccinia virus can use this pathway to enter host cells. Arf6-mediated endocytosis pathway involves the small GTPase Arf6 and is independent of both clathrin and caveolin. It has been implicated in the entry of viruses such as human papillomavirus (HPV) and herpes simplex virus (HSV). Cdc42-mediated endocytosis pathway involves the small GTPase Cdc42 and is also independent of clathrin and caveolin. It has been implicated in the entry of viruses such as coxsackievirus and adenovirus. Tyrosine kinase-dependent endocytosis pathway involves the activation of tyrosine kinases, such as epidermal growth factor receptor, and is dependent on DNM2. It has been implicated in the entry of viruses such as SV40 and adenovirus.

8.5 CNS INVASION OF RNA VIRUSES

The central nervous system (CNS) is protected by a barrier known as the blood–brain barrier (BBB), which is made up of brain microvascular endothelial cells (BMECs) joined by tight junctions and adheren junctions and surrounded by astrocytes and pericytes. The BBB selectively allows certain substances to pass through while restricting others, making it difficult for pathogens to invade the CNS. However, several viruses can enter the bloodstream and spread systemically throughout the body, including the CNS. To cross the BBB and enter the CNS, viruses must find a way to breach the barrier. One way this can happen is through the disruption of the tight junctions between BMECs, which can be caused by viral infection or inflammation. Alternatively, viruses can enter BMECs through various mechanisms, such as receptor-mediated endocytosis, and then use the cells as a Trojan horse to cross the BBB.

Another barrier that arboviruses must overcome to enter the CNS is the blood–cerebrospinal fluid barrier (BCSFB), which separates the blood from the cerebrospinal fluid. The BCSFB is made up of epithelial cells that form tight junctions, similar to the BMECs of the BBB, and restricts the passage of certain molecules. Viruses can potentially cross the BCSFB through infected cells or through the intercellular space between cells.

8.6 PATHWAYS EMPLOYED BY THE RNA VIRUS FOR CNS INVASION

8.6.1 Passive Diffusion

Viruses can simply passively diffuse between endothelial cells. This occurs when the endothelium is loose or injured, or when it is permeabilized by a virus or some other factor. Passive diffusion is generally not an efficient mechanism for virus entry, as it requires a large enough gap between endothelial cells for the virus to pass through. Viral infection or the release of inflammatory molecules can trigger the production of cytokines and chemokines that compromise the integrity of the tight junctions, allowing viruses to cross the barrier. This disruption can occur through various mechanisms, including the activation of signaling pathways that modify the cytoskeletal structure of the BMECs or the downregulation of proteins involved in maintaining tight junction integrity. Japanese encephalitis virus (JEV) is a mosquito-borne flavivirus that can cause severe encephalitis in humans. JEV infection induces the production of proinflammatory cytokines and chemokines, leading to the disruption of tight junction proteins and increased permeability of the BBB (Al-Obaidi et al. 2017; Li et al. 2015; Patabendige et al. 2018). West Nile virus (WNV), Zika virus, and measles virus infections have also been associated with increased BBB permeability, primarily mediated by the disruption of tight junctions between brain endothelial cells (Alimonti et al. 2018; Andrews 2010; Bhagat et al. 2021; Clé et al. 2020; Kaur et al. 2023; Papa et al. 2017).

8.6.2 Endothelial Cell Infection

Some viruses can directly infect and replicate within endothelial cells by employing specific host receptors, allowing them to release infectious viral particles on the basolateral side of the endothelium (Figure 8.2). This enables the virus to cross the endothelial barrier and infect surrounding tissues. There are various receptors on endothelial cells that RNA viruses can exploit for neuroinvasion.

LDL receptor-related protein 1 (LRP1) is a receptor expressed on endothelial cells that plays a role in the clearance of lipoproteins and other molecules from the bloodstream. It has been implicated in the entry of several RNA viruses into endothelial cells, including WNV, JEV, and dengue virus (DENV). These viruses can interact with LRP1 to facilitate their entry into endothelial cells and subsequent dissemination to the CNS.

AXL receptor tyrosine kinase on endothelial cells and has been identified as a receptor for Zika virus. Zika virus can bind to AXL on endothelial cells, leading to viral entry and potential neuroinvasion. Junctional adhesion molecule A (JAM-A) tight junction-associated protein is expressed on endothelial cells. It has been shown to be involved in the entry of certain RNA viruses, such as reovirus and coxsackievirus, into endothelial cells. Disruption of tight junctions or interaction with JAM-A can facilitate viral entry and subsequent dissemination to the CNS. Scavenger receptor class B type I (SR-BI) is a receptor involved in the uptake of lipoproteins and lipid transport. It has been implicated in the entry of flaviviruses, including WNV and DENV, into endothelial cells. These viruses can interact with SR-BI to gain entry into endothelial cells and potentially facilitate neuroinvasion. Mannose receptor (MR) is a pattern recognition receptor expressed on endothelial cells and has been associated with the entry of

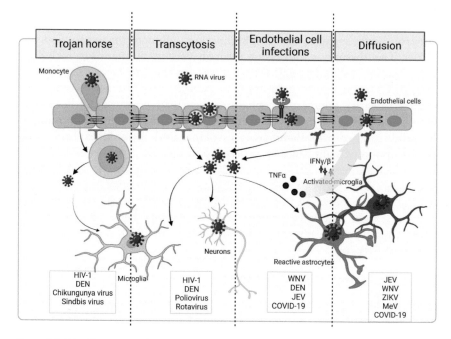

Figure 8.2 The illustration shows the blood–brain barrier and depicts four nonexclusive ways that RNA viruses can use to reach adjacent tissues. (1) *Diffusion*: When the integrity of the endothelium is compromised, viruses can freely diffuse across the barrier and enter the surrounding tissues. (2) *Infection*: The viruses infect the endothelial cells themselves. Once inside the cells, the viruses replicate and produce new viral particles, which are then released on the other side of the endothelium, allowing them to infect the adjacent tissues. (3) *Transcytosis*: Viruses can be taken up by the endothelial cells through endocytosis. Instead of being degraded, they are transported across the cell and then exocytosed on the other side of the endothelial barrier. (4) *Cell transport* (Trojan horse method): Some viruses can hijack leukocytes (immune cells) as a means of transport. The viruses can infect or attach themselves to these circulating leukocytes, which then carry the viruses through the endothelial cell wall, allowing them to bypass the barrier and enter the adjacent tissues.

certain viruses, such as HIV and SARS-CoV, into endothelial cells. While the neuroinvasive potential of these viruses may vary, their interaction with MR on endothelial cells can contribute to their entry and dissemination. It is important to note that the specific receptors involved in neuroinvasion can vary among different RNA viruses, and our understanding of these mechanisms continues to evolve. Further research is needed to fully elucidate the receptor usage and entry pathways employed by RNA viruses for neuroinvasion.

8.6.3 Virus Transcytosis

Virus transcytosis is a mechanism employed by certain viruses to cross the endothelial barrier and infect neighboring tissues. In this process, the viruses are internalized by endothelial cells through endocytic pathways but manage to evade degradation by the lysosomal pathway. Instead, they are transported within non-degradative endosomal vesicles across the cytoplasm of the endothelial cell. This transcellular transport allows the viruses to pass through the cell and be released on the opposite side of the endothelial barrier.

Transcytosis provides a means for viruses to traverse the endothelial barrier without compromising its integrity. By utilizing this mechanism, viruses can effectively disseminate to target tissues and organs, including the CNS. Some examples of viruses that utilize transcytosis for endothelial cell crossing and subsequent tissue infection include DENV and HSV. During transcytosis, the viruses exploit the cellular machinery of the endothelial cells to facilitate their transport. They may interact with specific receptors or molecules on the cell surface that trigger internalization and subsequent trafficking within the endosomal system. Fc receptors are a group of cell surface receptors that bind to the Fc region of antibodies. They are involved in immune responses and can facilitate the transport of immune complexes across endothelial cells. Some viruses, such as HIV, have been shown to exploit Fc receptors for transcytosis across endothelial barriers. LRP1 receptors have been implicated in the transcytosis of several viruses, including HSV and reovirus. LRP1 can interact with viral particles or viral envelope proteins, facilitating their transport across endothelial cells. Transferrin receptor is involved in the cellular uptake of transferrin-bound iron. Some viruses, such as picornaviruses and adenoviruses, can exploit transferrin receptor-mediated endocytosis for transcytosis across endothelial barriers. The precise mechanisms underlying transcytosis and the interactions between viruses and endothelial cells are still being investigated. Understanding the mechanisms of virus transcytosis can provide insights into viral pathogenesis and the development of targeted interventions. By identifying key receptors or cellular factors involved in this process, it may be possible to design strategies to disrupt virus transport and prevent the spread of infection to vital tissues and organs.

8.6.4 Cell-Associated Virus Transport

Some viruses can infect or bind to blood cells and use them as a Trojan horse to cross the endothelial barrier. The infected cells migrate through the endothelial cells via either paracellular or transcellular migration and release the virus into the surrounding tissue. This mechanism is used by some viruses, such as HIV, which can infect CD4+ T cells and use them to cross the BBB and infect the brain.

The passage of viruses through the BBB and into the CNS is a complex process that can occur through various mechanisms. For JEV and WNV, it is believed that virions initially enter the CNS through an intact BBB and subsequently disrupt the BBB through inflammatory signals. In contrast, for DENV, it has been shown that the virus can induce vascular permeability through the flavivirus NS1 protein, which leads to increased passive diffusion of the virus to the BBB.

The Trojan horse strategy involves viruses infecting or being carried by blood-circulating cells, which then cross the BBB through paracellular or transcellular migration. It is difficult to determine whether the immune infected cells found in the brain correspond to the first wave of invasion or to secondary infiltrations responding to an already established brain infection. It is possible that inflammation plays a significant role in viral BBB crossing and in the worsening of virus-induced neuropathology, but it may not be critical for initial cerebral colonization. To better understand the sequence of events that occur during viral neuroinvasion, BBB leakage, and inflammation, it is important to study the timing of these events in greater detail using relevant in vivo or ex vivo models.

BIBLIOGRAPHY

Al-Obaidi, M. M. J., et al. Japanese encephalitis virus disrupts blood-brain barrier and modulates apoptosis proteins in THBMEC cells. *Virus Research*, 2017, Apr 2;233:17–28. doi: 10.1016/j.virusres.2017.02.012. Epub 2017 Mar 6. PMID: 28279803.

Alimonti, J. B., et al. Zika virus crosses an in vitro human blood-brain barrier model. *Fluids and Barriers of the CNS*, 2018, May 15;15(1):15. doi: 10.1186/s12987-018-0100-y.

Andrews, S. FastQC—A quality control tool for high throughput sequence data. *Babraham Bioinformatics*, 2010. http://www.Bioinformatics.Babraham.Ac.Uk/Projects/Fastqc/

Bhagat, R., Kaur, G., Seth, P. Molecular mechanisms of Zika virus pathogenesis: An update. *Indian Journal of Medical Research*, 2021, Mar;154(3):433–445. doi: 10.4103/ijmr.IJMR_169_20. PMID: 35345069.

Clé, M., et al. Zika virus infection promotes local inflammation, cell adhesion molecule upregulation, and leukocyte recruitment at the blood-brain barrier. *mBio*, 2020, Aug 4;11(4):e01183-20. doi: 10.1128/mBio.01183-20.

Kaur, G., Pant, P., Bhagat, R., Seth, P. Zika virus E protein alters blood-brain barrier by modulating brain microvascular endothelial cell and astrocyte functions. *bioRxiv*, 2023. 10.1101/2023.02.09.527854. http://biorxiv.org/content/early/2023/02/10/2023.02.09.527854.abstract.

Li, F., et al. Viral infection of the central nervous system and neuroinflammation precede blood-brain barrier disruption during Japanese encephalitis virus infection. *Journal of Virology*, 2015, May;89(10):5602–5614. doi: 10.1128/JVI.00143-15. Epub 2015 Mar 11. PMID: 25762733.

Papa, Michelle P., et al. Zika virus infects, activates, and crosses brain microvascular endothelial cells, without barrier disruption. *Frontiers in Microbiology*, 2017, Dec 22;8:2557. doi: 10.3389/fmicb.2017.02557. PMID: 29312238.

Patabendige, Adjanie, Benedict D. Michael, Alister G. Craig, and Tom Solomon. Brain microvascular endothelial-astrocyte cell responses following Japanese encephalitis virus infection in an in vitro human blood-brain barrier model. *Molecular and Cellular Neuroscience*, 2018, Jun;89:60–70. doi: 10.1016/j.mcn.2018.04.002. Epub 2018 Apr 7. PMID: 29635016.

Rampersad, Sephra, and Paula Tennant. 2018. Replication and expression strategies of viruses. *Viruses: Molecular Biology, Host Interactions, and Applications to Biotechnology*, 2018, 55–82. doi: 10.1016/B978-0-12-811257-1.00003-6. Epub 2018 Mar 30.

Riezman, Howard, Philip G. Woodman, Gerrit Van Meer, and Mark Marsh. Molecular mechanisms of endocytosis. *Cell*, 1997; 91(6):731–738. ISSN 0092-8674, https://doi.org/10.1016/S0092-8674(00)80461-4

Part III

TECHNOLOGICAL AND THERAPEUTIC STRATEGIES

Chapter 9

RNA Sequencing for Characterizing Virus-Induced Changes

Conor J. Cremin

9.1 ANALYZING THE TRANSCRIPTOMIC LANDSCAPE

The central nervous system (CNS) is a highly sophisticated network comprising hundreds of different neural cells and other supportive cell types to regulate the function of all biological systems within a host. Cell types of the CNS have been shown to operate a wide range of different functions. With such a high degree of heterogenicity between these cell types, it is important to understand that different cell types may respond differently to the same stimulus. It is therefore crucial for transcriptomic studies of the CNS to enable the correct identification of responsive cell types with viral-induced changes in their transcriptome.

Many viruses have been identified to be able to infect CNS tissues through several different mechanisms (1). Several viruses can infiltrate into connective tissues to access tissues of the peripheral nervous system and jump across the terminal ends of sensory neurons (Figure 9.1A–C). Through the migration of viral progeny via axonal transport processes (2), these viruses can infiltrate into the tissues of the CNS. Many immunosuppressive viruses can travel across the blood–brain barrier (BBB) through a "Trojan horse" mechanism by infecting immune cells that subsequently pass through the BBB, which enables their entry to the brain (3) (Figure 9.1D). Others may have less defined entry mechanisms potentially through a complex paracellular or transcellular transportation process (4) (Figure 9.1E).

In analyzing the interactions between an invading virus and the infected CNS tissues of a host, it is important to have a clear understanding as to what tissues the virus wants to migrate into. Such sites of infection provide the most beneficial environment for the virus to potentially replicate and to

DOI: 10.1201/9781003285823-9

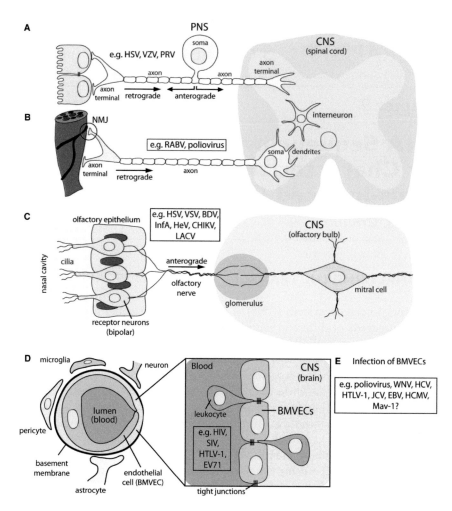

Figure 9.1 Mechanisms for viral entry into the CNS. (A) Alpha herpesviruses such as HSV-1 and VSV are able to propagate and travel via anterograde axonal transport from sensory neurons in the peripheral nervous system to the spinal column, which can result in lifelong infections. (B) The RABV and poliovirus can jump across neuromuscular junctions and propagate in somatic motor neurons up to the spinal cord. (C) Several viruses can infect the olfactory epithelium in the nasal passages and travel along the olfactory nerve into the brain. (D) Infiltration through the BBB. The BBB is composed of brain microvascular endothelium cells (BMVECs) with specialized tight junctions, surrounding basement membrane, pericytes, astrocytes, and neurons. Infected leukocytes can traverse this barrier carrying virus into the brain parenchyma. (E) Alternatively, virus particles in the bloodstream can infect BMVECs, compromising the BBB.

Source: This figure was taken from (1). All permissions for use in this chapter were obtained.

hijack neural cell processes to aid long-term survival of the virus and develop pathogenesis within an infected host.

Transcriptomic studies of infected tissues have identified that viruses can interfere with the host cell signaling processes that control the functionality of many neural cells (3,5,6). Such interference causes disruption to the cell's ability to regulate gene expression, resulting in a different expression profile within infected cells in comparison to uninfected cells of the same cell type. The purpose for viruses to induce changes to the transcriptome of infected cells depends on the nature of the specific virus. Many transcriptomic studies have analyzed how infected hosts respond to viral infection and have often showed that much of the induced gene expression in response to viral infection is initially to eliminate the virus from the infected cell population. An example of this can been seen in how human neural progenitor cells can induce the expression of AGO2 components to enable the cleavage of the RNA genome of the invading Zika virus (7). However, the opportunistic nature of most highly virulent viruses would give the expectation that some viral-host interactions would result in viral-induced changes that benefit the virus at some point during its life cycle. One of the best studied examples is the West Nile virus, which can cross the BBB and induce the development of neurodegenerative disorders in the hippocampus (8). Its viral components have been identified to interfere with several signaling which control the expression of components that regulate ubiquitin-signaled protein degradation and the expression of immune genes, IL-1β and IFN-γ (9). Viral-host interactions in neural cells of the CNS have also been linked at preserving viruses within neural cells, which can lead to lifelong infections (10). Although much is known about the types of viral-host interactions and their effects on the life cycle of neurotropic viruses, there is still much left unknown has to how the transcriptomic environment changes during viral infection.

Before next-generation sequencing technologies can be used to analyze viral-induced alterations to a neural cell transcriptome, it is important to be able to isolate the correct cellular population from the infected host. Apart from cells that are infected directly by the virus, infection will also trigger neural cells to recruit immune cells to the location of infection (11). Careful consideration must be given to the design of experiments when analyzing viral-induced changes in the transcriptome of the CNS as changes may be either directly or indirectly caused by infection due to the recruitment of immune cells by infected neural cells (11). In these cases, if a general sampling of cells extracted from the CNS was taken, characterization of viral-induced transcriptomic changes may include broad changes as a result of the host immune system. Although these changes are interesting for analyzing a host's response to infection, it is important to be aware that these changes may not be specific to a particular virus and can occur in other neurological disorders. Clarity on the exact type of viral-induced changes and their specificity is key when selecting the desired cell population for transcriptomic studies. Several methods exist for the specific extraction of select cells. Microdissection can enable the removal to small cell clusters from specific regions of a host's CNS (Figure 9.2A). This can be done with a laser or high-precision scalpel. Specific cell types can have unique surface biomarkers that can be used to identify these cells from a cell suspension of mixed cell types. When incubated with an antibody that is complementary to these cell type-specific biomarkers, these cells can be extracted from the mixed cell suspension (Figure 9.2B). Fluorescence-activated cell sorting can distinguish multiple cell populations, which can be an advantage for sequencing regions with large cellular populations of mixed cell types (12,13) (Figure 9.2C). For cells that are difficult to define or isolate through other means, several studies have genetically modified biomarkers to include a tag protein to enable an easier extraction process for these cells (Figure 9.2D). Once the ideal cell types have been isolated, the RNA material in these cells can be extracted and sequenced for the characterization of viral-induced transcriptomic changes.

9.2 IMPACTS OF THE EMERGENCE OF NEW RNA-SEQ TECHNOLOGIES

The abundance of gene expression has been shown to have significant variation across different cell types and locations within the CNS. Neural cells of the CNS also have a higher frequency of alternative splicing in gene expression (14) in comparison to most cells. Due to the extent of this variation, there is a push in the development of sequencing technology for methods that can provide an accurate characterization of transcriptomes with a wide range of gene expression mechanisms and transcript isoforms (13). A comprehensive analysis of a cell's transcriptome will provide more understanding as to how these cells respond to stimuli such as viral infection but also to other pathogens.

Many of the early studies in characterizing the functions of the CNS relied on the use of microarray assays (15,16). However, it was shown that these approaches were not able to cope efficiently with analyzing complex transcriptomes such as those with non-coding gene expression and alternative splicing. Although these platforms are still a popular choice in gene expression studies, they have largely been superseded by RNA-seq approaches (13). RNA-seq has enabled genome-wide transcriptional analysis and in-depth studies into identifying unknown functions across many cell types (17,18). For studies of the CNS, RNA-seq has generated invaluable data that has allowed for cellular differentiation across neural cell types and has even enabled the detection of new cell types (19). Many of these findings stem from the detection of cell type-specific gene expression patterns from RNA-seq data. Functional analysis of RNA-seq data can also determine how different cells respond to disease.

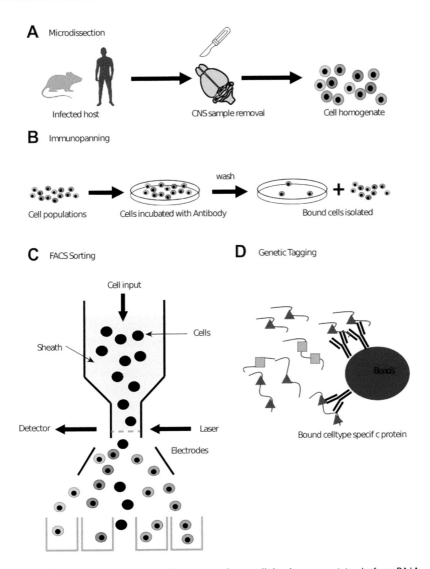

A Microdissection

Infected host   CNS sample removal   Cell homogenate

B Immunopanning

wash

Cell populations   Cells incubated with Antibody   Bound cells isolated

C FACS Sorting

Cell input

Cells

Sheath

Detector ← Laser

Electrodes

D Genetic Tagging

Beads

Bound celltype specif c protein

Figure 9.2 Sample extraction procedures to reduce cellular heterogenicity before RNA sequencing. (A) Microdissection can isolate small cell populations from the nervous tissue of a live host. (B) Immunopanning uses antibodies anchored to an inert surface which are complementary to the surface biomarkers for a specific cell. Washing enables the removal of unbound cells while targeted cells are separated from the mixed cell suspension. (C) Fluorescence-activated cell sorting can separate cells based on a number of different criteria depending on the objective. Cells can be sorted based on biomarker presence, granularity, size, or fluorescence. (D) Genetic tagging offers the extraction of cells with tag sequences inserted artificially using antibodies that are complementary of the tag sequences.

Neural cells have been shown to generate a significant number of transcript variants through their alternative splicing machinery (14,20). Furthermore, a significant proportion of the human genome in cells of the CNS encode non-coding genes. Functions for many of these novel transcripts are unknown but indicate a very complex and dynamic transcriptomic landscape. The identification of cell type-specific gene expression can also become further complicated when neural cells are exposed to pathogens. Cells of the CNS will invoke a host response to minimize neurological degradation of the surrounding tissues of the CNS (21). To differentiate between transcription that is specific to standard cell processes of select cell types and altered gene expression due to pathogens, like viruses, it is important to perform differential expression between infected cell populations and uninfected healthy controls.

Though studies that have used traditional bulk RNA-seq analysis have generated significant findings in identifying transcriptomic changes, these methods lack the ability to analyze changes across tissues with high heterogenicity across cells like those in the CNS (22,23). The emergence of single-cell RNA-seq methods enables researchers to detect viral-inducible transcriptomic changes at a cellular level. This circumvents many of the limitations imposed by bulk RNA-seq analysis and offers significant sensitivity in the detection of cell type specific gene expression. However, single-cell RNA-seq (scRNA-seq) pipelines have had to overcome several limitations in the characterization of neural tissues. Rare neural cell types with small populations can be difficult to characterize through traditional droplet-seq based methods which can require larger cell numbers to enable cell differentiation due to shorter read lengths generated from sequencing (24). This can make it difficult in the analysis for the occurrence of splicing and ncRNA gene expression in neural cells (25). More recent developments in sequencing technology have enabled the development of SMART-seq technology. This has enabled better analysis across a larger number of genes and with smaller cell populations due to the use of longer read sequences (26).

The analysis of a host's response to viral infection using RNA-seq has provided insight into the inner workings of a host's genome. Characterizations of gene expression and functional analysis have enabled many studies to assign functions to previously unknown regions of the genome. Many of these studies have emphasized the importance of non-coding RNA gene expression in modulating a host's response to viral infection (27,28). This has been shown to be a promising hypothesis in the analysis of several neurotropic viruses such as the rabies virus (29), Japanese encephalitis virus (JEV) (30), and the SARS-CoV-2 virus (31,32). There have been many studies that have proposed functions for ncRNA transcripts expressed in neural cells. Some have shown that several long non-coding RNA (lncRNA) genes can trigger neuroinflammation in CNS tissues (33). This can be a dangerous in CNS tissues and could lead to the development of other neurological disorders. Several studies have shown that severity of neuroinflammation is highly dependent on the release of cytokines from viral infected neural cells (34,35). Therefore, there is growing interest in exploring how changes in the transcriptome of infected neural cells can incite the activation of innate immune system.

9.3 A GLOBAL CONTEXT: INFERRING CHANGES TO HIGHER-ORDER STRUCTURES

Functional analysis of gene expression data can determine if collections of genes are shown to share properties or operate together as part of biological pathway within a cell. If a collection of functionally similar genes are shown to be expressed in response to a virus it is possible that viral infection has induced the activation of a specific signaling pathway within a host. This could highlight new insights into the regulation of antiviral defenses within a host or possible viral-specific mechanisms to antagonize the host's innate immune system (36). To characterize viral-host interactions, it is necessary to understand how the expression of responsive genes are regulated. Several consortiums have built software that enables the identification or prediction of many transcription factors and their accessory modulators for RNA-seq data such as DoRothEA (37,38), and oPOSSUM-3 (39). However, many neural cellular networks utilize multiple transcription factors for gene expression and are affected by other modulators which can also control the transcription process (40). Therefore, it is important to assess if virus-induced changes in gene expression are caused by alterations to higher-order structures in the nucleus of viral-infected neural cells, such as through the remodeling process of chromatin configurations and epigenetics.

The characterization of gene expression in response to viral infection from RNA-seq data can be used to identifying if collections of genes function as part of tightly controlled biological networks. By comparing the regulatory properties of genes, it is possible to identify if subsets of genes are controlled by the same epigenetic regulatory system. The role of epigenetics in the CNS has been studied in detail to play a role in neurodevelopment, neurogenesis, neural behavior, and the manifestation of neuropsychiatric disorders (41). Infection studies of live rat models have identified significant correlations between changes in DNA methylation profiles and gene expression of immune genes in neural cells located in specific cortexes of the brain (42). Such profiles have been closely matched to those of several neurodevelopmental disorders such as schizophrenia and bipolar disorder (43,44). These results have implicated that viral infection of neural cells can arrest the development of neurological processes in the CNS. Much of the neurological damage associated with viral infection seems to arise from an excessive activation of the immune response in infected hosts. Infection with the JEV in human microglial cells has been shown to cause neuroinflammation within the CNS. It was shown that JEV infection induced the expression of microRNAs which are known to regulated post-transcriptional gene expression processes (45). This resulted in the enrichment of pathways relating to cytokine signaling, Toll-like receptor signaling, WNT signaling, and apoptosis (46). The exact mechanism of how viral infection could alter DNA methylation processes within infected neural cells is still unclear but is likely due to the viral manipulation of cellular DNA methyltransferases (DMNTs). An example of this process can be seen with the Epstein–Barr virus (EBV), which is shown to regulate the function of DMNTs to switch between its lytic and latent states of infection (47,48). Through a better understanding of specific gene expression patterns, future analysis should be able to identify viral-induced epigenetic changes from RNA-seq datasets.

Histone modifications are also an important aspect of how neural cells can regulate gene expression through epigenetics. Histone deacetylases (HDACs) are known to remove acetyl groups from susceptible histone residues to enable mobility of nucleosome structures to expose or conceal genes from the surrounding transcriptomic machinery. Studies have shown that neural infection by the SARS-CoV-2 virus is heavily dependent on the function of HDACs (49). It has also been suggested that histone configurations could also be controlled in neural cells through an intricate network of non-coding RNA HDAC inhibitors (HDACi) (50). Therefore, there is a growing interest in the exploration of the therapeutic use of HDACis in the treatment and prevention of neurological disorders because of viral infection. Other mechanisms of the manipulation of histone configurations have been shown to involve the integration of viral genomes into the infected host's own genome. Both EBV and HIV infections have been linked to express a variety of different coding and non-coding components to regulate the functions of the infected host's HDACs and histone methyltransferases (HMTs), which ultimately can recruit histone units to package the viral genome (51–53). However, many of these mechanisms are very complex, and it is unclear if such mechanisms are used by these viruses across all the susceptible cell types of the CNS.

Since neural cells of the CNS are shown to have a high degree of alternative splicing, it stands to reason that the chromatin structure of the nucleus of neural cells must undergo a significant amount of remodeling to make room for the components of the spliceosome to facilitate efficient splicing (54). It has been shown that chromatin remodeling can play a key role in the maturation process of neural stem cells and is involved in the development of several key areas of the brain. Defects in this process have been linked to causing several neurodevelopmental disorders; however, its function in viral infected neural cells is less understood (55). Chromatin remodeling has been identified as an important factor in the regulation of host-cell processes in response to viral infection across many tissues (Figure 9.3).

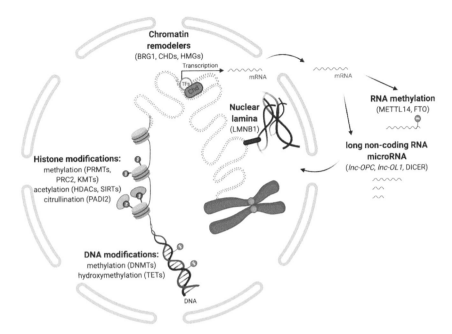

Figure 9.3 Chromatin remodeling and epigenetic regulation of a host's genome. Epigenetic regulation can take the form of DNA modifications, histone modification, or the interaction of lncRNA with transcriptomic machinery, all of which assist in the regulation of gene expression at specific locations along the genome. Chromatin remodeling, association with the nuclear lamina, and post-transcriptional regulation by RNA methylation and long non-coding and microRNA are also involved.

Source: From Pruvost M, Moyon S. Oligodendroglial epigenetics, from lineage specification to activity-dependent myelination. *Life (Basel)*. 2021;11(1):62. As MDPI is an open access journal, all rights and permissions were obtained under its open access license.

Recent studies have suggested that HIV can modulate chromatin configurations in neurons to control the expression of several gene components necessary for the TCA cycle. This can result in impaired cognitive function in infected individuals (56).

The analysis of RNA-seq data offers a great opportunity to infer how viral infection can alter epigenetic regulation within infected cells. However, it can be difficult to make direct connections between gene expression profiles and the status of epigenetic modifiers across large gene sets. Visualization of RNA-seq alignments using genome brewers such as the UCSC genome browser (57) and the Integrative Genomics Viewer (58) can provide insight into identifying changes to the expression coverage along genes of interest which may be due to viral interference with epigenetic modulation.

However, to undertake a more comprehensive analysis in characterizing how viral infection can cause significant changes in gene expression data, it is necessary to combine RNA-seq data with other omics datasets.

9.4 BRINGING IT ALL TOGETHER: DATA AGGREGATION

Multiomics is the analytical approach of utilizing multiple datasets across several of the omics disciplines to perform an integrative analysis with the ultimate objective of discovering new connections between biological concepts (59). The implementation of such approaches would enable a characterization of viral-induced changes across many molecular pathways that are responsible for the control of gene expression within infected cells. Popular examples of combinations of different omics datasets often include the use of ChIP-seq and RNA-seq datasets. Such studies often aim to identify if differential binding of a target protein correlates with differential gene expression. Based on the accumulation of omics datasets, several databases now exist which describe the genomic locations and cell types where certain protein-binding events are known to regulate specific gene expression within the CNS (60). Although multiomics is a powerful approach, it is only beginning to have frequent use as an analytic tool in the study of viral infection of the CNS. Limitations to its widespread use mostly stem from the lack of suitable different omic datasets derived from CNS tissues but also due to the ambiguity that is inherent in the identification of some neural cell types (61). However, recent progress is still being made to make multiomics more applicable to the study of neurotropic viruses (62).

With the abundance of RNA-seq datasets now available, large annotation databases such as CellMarker (63) are also being constructed from these collections to enable rapid identification of neural cell types. This is of particular importance for the analysis of viral-induced changes to the transcriptome of neural cells because many of the intrinsic signaling processes across the same cell types have been shown to vary based on their location within the CNS (64). This can create difficulty in deciphering true molecular changes induced by viral infection from normative cell signaling processes and can lead to misinterpretation of results.

Data aggregation of RNA-seq data can also be used to examine the reproducibility of gene expression signatures induced by viral infections. Many RNA-seq studies have identified key genes that are expressed in response to viral infection and are subsequently thought to have a significant role in the life cycle of viral infection (65). Viral infection experiments tend to have an array of parameters that can influence the extent of gene expression other than the viral infection process itself. Parameters such as infection time points, cell types sampled, reagents, cell topology, temperature, and even operator technique can induce non-specific sources of gene expression in RNA-seq data. A meta-analysis of RNA-seq datasets derived from similar viral infection studies could be performed to identify non-specific gene expression (66). Gene expression which is shown to occur in only a small number of RNA-seq studies could mean that expression is specific to the particular conditions of these studies and not part of a reliable response to viral infection.

Pattern recognition across RNA-seq data collections can offer new insights into the mechanisms of gene networks within select tissues of a host. Co-expression analysis has been to be a useful tool in collecting genes to functional networks within cells from gene expression data. Many studies have used co-expression to predict the operation of functions within many regions of the brain These co-expression networks can be used to predict how tissues may respond to disease. By identifying genes with strong co-expression from differential expression data, studies can show which genes operate as part of functional networks in neural tissues. Defining clusters of co-expressed genes could also be useful in identifying the function of unannotated genes. Several co-expression studies have had success in determining the function of genes during the development of neurodegenerative disorders (67–69). This has been particularly beneficial to the characterization of ncRNA genes which are shown to have an increasing significance in the development of several diseases in the CNS. Co-expression between genes can be difficult to interpret an exact biological meaning; however, they can be useful in characterizing the behavior of genes across different states of disease.

Machine learning (ML) can also infer clinical outcomes based on experimental models built from RNA-seq datasets. Many of these models are trained to determine the severity of disease pathogenesis by identifying the expression of specific biomarkers. A prediction can be made to determine the most probable prognosis (70). There are a number of different methods and statistical algorithms in building predictive models. Availability of RNA-seq data derived from different tissues of the CNS has now made it possible to explore disease pathogenesis within many regions of the CNS. Due to the time and computational demands required for the construction of large training models, it is becoming a popular choice to use pre-built models from earlier ML studies in the analysis of new RNA-seq data. Many models can also be updated or retrained with relative ease without the need of a complete reconstruction. However, this should only be undertaken as a last resort or when the quality of current datasets is better than those used previously.

9.5 CONCLUSIONS AND PERSPECTIVES

Viral infection of the CNS has been shown to cause a number of neurological disorders depending on the specific area where infection occurs. Infection has been shown to invoke a significant host response across many different cell types within the CNS. RNA-seq analysis has been shown to enable accurate characterization of a host's response to viral infection. However, due to the high degree of cellular heterogenicity with CNS tissues, it is important for RNA-seq analytic pipelines to be able to differentiate between different cell types within the CNS. Comprehensive analysis of RNA-seq data can identify regulation pathways induced by viral infection. Advancements in the recognition of specific patterns in gene expression can also pinpoint the involvement of epigenetic pathways in the control of a host's response to viral infection.

The amount of RNA-seq data available from public repositories enables studies to perform high-powered analysis in detecting reliable changes to a host's transcriptome caused by viral infection. Prediction studies using ML concepts on RNA-seq data collections are increasingly showing promise in determining accurate clinical outcomes for several diseases of the CNS. Although the current number of viral infection datasets of the CNS is small, it will be possible to adopt a more widespread use of ML methods in the near future.

REFERENCES

1. Koyuncu Orkide O, Hogue Ian B, Enquist Lynn W. Virus infections in the nervous system. *Cell Host Microbe* 2013;13(4); 379–393.
2. Gilman S, ed. Index. In *Neurobiology of Disease*. Burlington: Academic Press; 2007, 1047–1085. eBook ISBN: 9780080466385.
3. Chen Z, Li G. Immune response and blood–brain barrier dysfunction during viral neuroinvasion. *Innate Immunity* 2021;27(2); 109–117.
4. Michlmayr D, Andrade P, Gonzalez K, Balmaseda A, Harris E. CD14(+)CD16(+) monocytes are the main target of Zika virus infection in peripheral blood mononuclear cells in a paediatric study in Nicaragua. *Nature Microbiology*. 2017;2(11); 1462–1470.
5. Wouk J, Rechenchoski DZ, Rodrigues BCD, Ribelato EV, Faccin-Galhardi LC. Viral infections and their relationship to neurological disorders. *Archives of Virology* 2021;166(3); 733–753.
6. Chang CY, Wu CC, Wang JD, Li JR, Wang YY, Lin SY, et al. DHA attenuated Japanese encephalitis virus infection-induced neuroinflammation and neuronal cell death in cultured rat Neuron/glia. *Brain, Behavior, and Immunity* 2021;93; 194–205.
7. Xu YP, Qiu Y, Zhang B, Chen G, Chen Q, Wang M, et al. Zika virus infection induces RNAi-mediated antiviral immunity in human neural progenitors and brain organoids. *Cell Research* 2019;29(4); 265–273.
8. Madden K. West Nile virus infection and its neurological manifestations. *Clinical Medicine & Research* 2003;1(2); 145–150.
9. Fulton CDM, Beasley DWC, Bente DA, Dineley KT. Long-term, West Nile virus-induced neurological changes: A comparison of patients and rodent models. *Brain, Behavior, & Immunity—Health* 2020;7; 100105.
10. Mangold CA, Rathbun MM, Renner DW, Kuny CV, Szpara ML. Viral infection of human neurons triggers strain-specific differences in host neuronal and viral transcriptomes. *PLoS Pathogens* 2021;17(3); e1009441.
11. Ousman SS, Kubes P. Immune surveillance in the central nervous system. *Nature Neuroscience* 2012;15(8); 1096–101.
12. Schwarz JM. Using fluorescence activated cell sorting to examine cell-type-specific gene expression in rat brain tissue. *Journal of Visualized Experiments* 2015;99; e52537-e.
13. Dong X, You Y, Wu JQ. Building an RNA sequencing transcriptome of the central nervous system. *Neuroscientist* 2016;22(6); 579–592.
14. Su CH, Tarn WY. Alternative splicing in neurogenesis and brain development. *Frontiers in Molecular Biosciences* 2018;5; 12.

15. Bareyre FM, Schwab ME. Inflammation, degeneration and regeneration in the injured spinal cord: insights from DNA microarrays. *Trends in Neurosciences* 2003;26(10); 555–563.

16. Cahoy JD, Emery B, Kaushal A, Foo LC, Zamanian JL, Christopherson KS, et al. A transcriptome database for astrocytes, neurons, and oligodendrocytes: A new resource for understanding brain development and function. *Trends in Neurosciences* 2008;28(1); 264–278.

17. Snyder M, Gerstein M. Genomics. Defining genes in the genomics era. *Science* 2003;300(5617); 258–260.

18. Havilio M, Levanon EY, Lerman G, Kupiec M, Eisenberg E. Evidence for abundant transcription of non-coding regions in the *Saccharomyces cerevisiae* genome. *BMC Genomics* 2005;6; 93.

19. Voineagu I, Wang X, Johnston P, Lowe JK, Tian Y, Horvath S, et al. Transcriptomic analysis of autistic brain reveals convergent molecular pathology. *Nature* 2011;474(7351); 380–384.

20. Iijima T, Iijima Y, Witte H, Scheiffele P. Neuronal cell type-specific alternative splicing is regulated by the KH domain protein SLM1. *Journal of Cell Biology*. 2014;204(3); 331–342.

21. Griffiths MR, Gasque P, Neal JW. The regulation of the CNS innate immune response is vital for the restoration of tissue homeostasis (Repair) after acute brain injury: A brief review. *International Journal of Inflammation* 2010;2010; 151097.

22. Li Z, Tyler WA, Haydar TF. Lessons from single cell sequencing in CNS cell specification and function. *Current Opinion in Genetics and Development* 2020;65; 138–143.

23. Chen G, Ning B, Shi T. Single-Cell RNA-Seq technologies and related computational data analysis. *Frontiers in Genetics* 2019; 10.

24. Li Z, Tyler WA, Haydar TF. Lessons from single cell sequencing in CNS cell specification and function. *Current Opinion in Genetics & Development* 2020;65; 138–143.

25. Arzalluz-Luque Á, Conesa A. Single-cell RNAseq for the study of isoforms—How is that possible? *Genome Biology* 2018;19(1); 110.

26. Cuevas-Diaz Duran R, Wei H, Wu JQ. Single-cell RNA-sequencing of the brain. *Clinical and Translational Medicine* 2017;6(1); 20.

27. Mandhana R, Horvath CM. Sendai virus infection induces expression of novel RNAs in human cells. *Scientific Reports* 2018;8(1); 16815.

28. Xu X, Mann M, Qiao D, Brasier AR. Alternative mRNA processing of innate response pathways in respiratory syncytial virus (RSV) Infection. *Viruses* 2021;13(2); 218.

29. Zhao W, Su J, Wang N, Zhao N, Su S. Expression profiling and bioinformatics analysis of CircRNA in Mice Brain infected with rabies virus. *International Journal of Molecular Sciences* 2021;22(12); 6537.

30. Zhou X, Yuan Q, Zhang C, Dai Z, Du C, Wang H, et al. Inhibition of Japanese encephalitis virus proliferation by long non-coding RNA SUSAJ1 in PK-15 cells. *Journal of Virology* 2021;18(1); 29.

31. Yang Q, Lin F, Wang Y, Zeng M, Luo M. Long Noncoding RNAs as emerging regulators of COVID-19. *Frontiers in Immunology* 2021; 12.

32. Meydan C, Madrer N, Soreq H. The neat dance of COVID-19: NEAT1, DANCR, and co-modulated cholinergic RNAs link to inflammation. *Frontiers in Immunology* 2020;11.

33. Chen Z, Wu H, Zhang M. Long non-coding RNA: An underlying bridge linking neuroinflammation and central nervous system diseases. *Neurochemistry International* 2021;148; 105101.

34. Wang G, Zhang J, Li W, Xin G, Su Y, Gao Y, et al. Apoptosis and proinflammatory cytokine responses of primary mouse microglia and astrocytes induced by human H1N1 and avian H5N1 influenza viruses. *Cellular & Molecular Immunology* 2008;5(2); 113–20.

35. van Riel D, Verdijk R, Kuiken T. The olfactory nerve: A shortcut for influenza and other viral diseases into the central nervous system. *Journal of Pathology* 2015;235(2); 277–287.

36. Bercovich-Kinori A, Tai J, Gelbart IA, Shitrit A, Ben-Moshe S, Drori Y, et al. A systematic view on influenza induced host shutoff. *Elife* 2016;5; e18311.

37. Holland CH, Tanevski J, Perales-Patón J, Gleixner J, Kumar MP, Mereu E, et al. Robustness and applicability of transcription factor and pathway analysis tools on single-cell RNA-seq data. *Genome Biology* 2020;21(1); 36.

38. Garcia-Alonso L, Holland CH, Ibrahim MM, Turei D, Saez-Rodriguez J. Benchmark and integration of resources for the estimation of human transcription factor activities. *Genome Research* 2019;29(8); 1363–1375.

39. Kwon AT, Arenillas DJ, Worsley Hunt R, Wasserman WW. oPOSSUM-3: Advanced analysis of regulatory motif over-representation across genes or ChIP-Seq datasets. *G3 (Bethesda)* 2012;2(9); 987–1002.

40. Schochet T, Lovatt D, Eberwine J. Chapter 27 — Transcription factors in the central nervous system. In Brady ST, Siegel GJ, Albers RW, Price DL, eds. *Basic Neurochemistry*, 8th ed. New York: Academic Press; 2012, 514–530.

41. Ravi B, Kannan M. Epigenetics in the nervous system: An overview of its essential role. *Indian Journal of Human Genetics* 2013;19(4); 384–391.

42. Richetto J, Massart R, Weber-Stadlbauer U, Szyf M, Riva MA, Meyer U. Genome-wide dna methylation changes in a mouse model of infection-mediated neurodevelopmental disorders. *Biological Psychiatry* 2017;81(3); 265–276.

43. Dong E, Dzitoyeva SG, Matrisciano F, Tueting P, Grayson DR, Guidotti A. Brain-derived neurotrophic factor epigenetic modifications associated with schizophrenia-like phenotype induced by prenatal stress in mice. *Biological Psychiatry* 2015;77(6); 589–596.

44. Kneeland RE, Fatemi SH. Viral infection, inflammation and schizophrenia. *Progress in Neuro-Psychopharmacology and Biological Psychiatry* 2013;42; 35–48.

45. Cao DD, Li L, Chan WY. MicroRNAs: Key regulators in the central nervous system and their implication in neurological diseases. *International Journal of Molecular Sciences* 2016;17(6); 842.

46. Rastogi M, Srivastava N, Singh SK. Exploitation of microRNAs by Japanese encephalitis virus in human microglial cells. *Journal of Medical Virology* 2018;90(4); 648–654.

47. Kalla M, Schmeinck A, Bergbauer M, Pich D, Hammerschmidt W. AP-1 homolog BZLF1 of Epstein-Barr virus has two essential functions dependent on the epigenetic state of the viral genome. *Proceedings of the National Academy of Sciences of the United States of America* 2010;107(2); 850–855.

48. Woellmer A, Arteaga-Salas JM, Hammerschmidt W. BZLF1 governs CpG-methylated chromatin of Epstein-Barr virus reversing epigenetic repression. *PLoS Pathogens* 2012;8(9); e1002902.

49. Teodori L, Sestili P, Madiai V, Coppari S, Fraternale D, Rocchi MBL, et al. MicroRNAs bioinformatics analyses identifying HDAC pathway as a putative target for existing anti-COVID-19 therapeutics. *Frontiers in Pharmacology* 2020;11; 582003.

50. Kumar S, Shanker OR, Kumari N, Tripathi M, Chandra PS, Dixit AB, et al. Neuromodulatory effects of SARS-CoV2 infection: Possible therapeutic targets. *Expert Opinion on Therapeutic Targets* 2021;25(6); 509–519.

51. Guo R, Gewurz BE. Epigenetic control of the Epstein-Barr lifecycle. *Current Opinion in Virology* 2022;52; 78–88.

52. Machida S, Depierre D, Chen H-C, Thenin-Houssier S, Petitjean G, Doyen CM, et al. Exploring histone loading on HIV DNA reveals a dynamic nucleosome positioning between unintegrated and integrated viral genome. *Proceedings of the National Academy of Sciences* 2020;117(12); 6822–6830.

53. Michieletto D, Lusic M, Marenduzzo D, Orlandini E. Physical principles of retroviral integration in the human genome. *Nature Communications* 2019;10(1); 575.

54. Lim DA, Suárez-Fariñas M, Naef F, Hacker CR, Menn B, Takebayashi H, et al. In vivo transcriptional profile analysis reveals RNA splicing and chromatin remodeling as prominent processes for adult neurogenesis. *Molecular and Cellular Neuroscience* 2006;31(1); 131–48.

55. D'Souza L, Channakkar AS, Muralidharan B. Chromatin remodelling complexes in cerebral cortex development and neurodevelopmental disorders. *Neurochemistry International* 2021;147; 105055.

56. Sanna PP, Fu Y, Masliah E, Lefebvre C, Repunte-Canonigo V. Central nervous system (CNS) transcriptomic correlates of human immunodeficiency virus (HIV) brain RNA load in HIV-infected individuals. *Scientific Reports* 2021;11(1); 12176.

57. Kent WJ, Sugnet CW, Furey TS, Roskin KM, Pringle TH, Zahler AM, et al. The human genome browser at UCSC. *Genome Research* 2002;12(6); 996–1006.

58. Robinson JT, Thorvaldsdóttir H, Winckler W, Guttman M, Lander ES, Getz G, et al. Integrative genomics viewer. *Nature Biotechnology* 2011;29(1); 24–26.

59. Hasin Y, Seldin M, Lusis A. Multi-omics approaches to disease. *Genome Biology* 2017;18(1); 83.

60. Zerbino DR, Wilder SP, Johnson N, Juettemann T, Flicek PR. The ensembl regulatory build. *Genome Biology* [Internet] 2015;16; 56. Available from: http://europepmc.org/abstract/MED/25887522; https://doi.org/10.1186/s13059-015-0621-5; https://europepmc.org/articles/PMC4407537; https://europepmc.org/articles/PMC4407537?pdf=render.

61. Zeng H, Sanes JR. Neuronal cell-type classification: Challenges, opportunities and the path forward. *Nature Reviews Neuroscience* 2017;18(9); 530–546.

62. Tedeschi A, Popovich PG. The application of omics technologies to study axon regeneration and CNS repair. *F1000Research* 2019;8; F1000 Faculty Rev-311.

63. Zhang X, Lan Y, Xu J, Quan F, Zhao E, Deng C, et al. CellMarker: A manually curated resource of cell markers in human and mouse. *Nucleic Acids Research* 2018;47(D1); D721–D728.

64. McKenzie AT, Wang M, Hauberg ME, Fullard JF, Kozlenkov A, Keenan A, et al. Brain cell type specific gene expression and co-expression network architectures. *Scientific Reports* 2018;8(1); 8868.

65. Cho H, Proll SC, Szretter KJ, Katze MG, Gale M, Diamond MS. Differential innate immune response programs in neuronal subtypes determine susceptibility to infection in the brain by positive-stranded RNA viruses. *Nature Medicine* 2013;19(4); 458–464.

66. Cremin CJ, Dash S, Huang X. Big data: Historic advances and emerging trends in biomedical research. *Current Research in Biotechnology* 2022;4; 138–151.

67. Lancour D, Dupuis J, Mayeux R, Haines JL, Pericak-Vance MA, Schellenberg GC, et al. Analysis of brain region-specific co-expression networks reveals clustering of established and novel genes associated with Alzheimer disease. *Alzheimer's Research & Therapy* 2020;12(1); 103.

68. Gerring ZF, Gamazon ER, Derks EM, for the Major Depressive Disorder Working Group of the Psychiatric Genomics C. A gene co-expression network-based analysis of multiple brain tissues reveals novel genes and molecular pathways underlying major depression. *PLoS Genetics* 2019;15(7); e1008245.

69. Hartl CL, Ramaswami G, Pembroke WG, Muller S, Pintacuda G, Saha A, et al. Coexpression network architecture reveals the brain-wide and multiregional basis of disease susceptibility. *Nature Neuroscience* 2021;24(9); 1313–1323.

70. Yu T-H, Su B-H, Battalora LC, Liu S, Tseng YJ. Ensemble modeling with machine learning and deep learning to provide interpretable generalized rules for classifying CNS drugs with high prediction power. *Briefings in Bioinformatics* 2021;23(1).

Chapter 10

Novel Methods and Technologies for Efficient Detection of RNA Viruses in the Human Brain

Sukh Sandan, Ridhika Bangotra, Parveen Kumar, Neha Goel, and Mohit Sharma

10.1 INTRODUCTION

Viruses are pervasive entities that require a host to replicate. They can infect a wide variety of life types, ranging from bacteria to plants and animals, and they can be found in all ecosystems (Sime-Ngando 2014). Viral diseases have a tremendous negative impact on the world's economy and health. Interactions between humans and animals are the main cause of the spread of existing viruses and the generation of new ones (Zhao et al. 2019). Climate change, social activity, intensive farming, and globalization are additional variables influencing the spread of new viruses (Trubl et al. 2020). The genome of these viruses consists of nucleic acids (either double- or single-stranded DNA or RNA) and a capsid which is made up of several copies of a few specific proteins. RNA viruses constitute one of the major classes of pathogenic organisms causing human diseases, with varying degrees of severity (Cassedy et al. 2021; Trubl et al. 2020). The major problem associated with RNA viruses is their quick adaptation to the rapidly changing environment, which leads to their genetic mutation. The mutation rate of RNA viruses is up to a million times higher than that of their hosts, and they can incorporate mutated nucleotides at a rate of 10^4 to 10^6 substitutions per nucleotide per infected cell (Pachetti et al. 2020). The ability of viruses to quickly mutate and spread between hosts makes the development of dynamic approaches to identify existing and emerging viral strains imperative (Yang et al. 2022). Taking into consideration the severity of RNA viruses and diseases associated with them, rapid diagnosis is crucial for selecting appropriate preventative and therapeutic measures.

There are various methods available for the detection of viruses. Conventional methods focus on morphological characteristics observed by light microscopy or transmission electron microscopy in a variety of specimens, including cell cultures and fertilized eggs. Also, it is feasible to identify viruses, sometimes down to the species level, through serology and antibody-based diagnostics (Shifman et al. 2019). But these methods primarily lack strain specificity and just provide morphological cues. Currently, the most widely

DOI: 10.1201/9781003285823-10

used techniques for promptly detecting viral infection include nucleic acid–based detection (polymerase chain reaction [PCR], reverse transcription PCR, and microarrays) and immunoassay techniques (Cassedy et al. 2021). High sensitivity is typically provided by nucleic acid-based detection, but it can also be time-consuming, expensive, and labor-intensive. Immunoassays, on the other hand, provide robustness at a lower cost. Additionally, certain immunoassay formats that employ lateral-flow technology can produce results relatively quickly (Cassedy et al. 2021; Su et al. 2021). Immunoassays, however, often are unable to attain comparable sensitivity as compared to nucleic acid–based detection techniques (Su et al. 2021). However, the emergence of next-generation sequencing (NGS) in diagnostics can produce extremely precise results. The ability of NGS to sequence a large number of viral genomes would benefit researchers in tracking infections in a short time period. Additionally, NGS enables the direct detection, identification, and discovery of viruses in an unbiased manner without requiring antibodies or prior knowledge of the pathogen macromolecular sequence (Shifman et al. 2019). High-throughput sequencing (HTS) technologies have the potential for comprehensive pathogen detection (unknown samples) without any prior assumptions about the features of the organisms. Another benefit of this technology is the capability to identify non-culturable organisms, co-infections, drug resistance, and others (Shifman et al. 2019).

There is increased prevalence of neurological diseases worldwide (Fullard et al. 2021). The most frequent etiologies of central nervous system (CNS) infections are viral, and the cause of about 50% of brain infections is still unknown. The most common cause of meningitis, encephalitis, and meningoencephalitis is virus, while other pathogens can also cause these diseases (Zanella et al. 2019). In pediatric and adult populations, enterovirus, herpes simplex types 1 and 2 (HSV-1 and HSV-2), and varicella zoster virus are the viruses most frequently linked to encephalitis, meningoencephalitis, and meningitis (Swanson and McGavern 2015). The incidence of other viruses varies depends on the patient's immune system and geographic area. However, HTS has made it possible to objectively examine viral etiologies, opening up new research directions in the study of CNS infections (Shifman et al. 2019). Therefore, this chapter focuses on various novel/efficient methods and techniques for the detection of RNA viruses affecting the human nervous system/brain.

10.2 HIGH-THROUGHPUT SEQUENCING IN VIRUS DETECTION

The development of HTS-based techniques has opened the door to pathogen identification using genomics without making any assumptions about the characteristics of the organisms (Hilton et al. 2016). The high ratio between the genome sizes of hosts and infections limits the utility of HTS for the characterization of unknown viral viruses in pertinent clinical samples. This restriction is particularly relevant to blood-related samples because of the high white blood cell content. However, the application of HTS for virus identification in clinical samples is increasing (Shifman et al. 2019). HTS has been conducted primarily in less relevant clinical samples (e.g., brain) and provided diagnoses in shorter time frames because they did not require culture first, while other methods obtained tens to a few thousand viral reads that might be sufficient for sequencing against a model organism but not for characterizing novel viruses. Shifman et al. (2019) reported the discovery of a new species of Orthobunyavirus that was isolated from horse plasma. The identification was made using a straightforward HTS methodology. After enrichment in cell culture, RNA was isolated from the growing medium, and in less than 12 hours rapid library preparation, HTS, and primary bioinformatic analyses were completed. The taxonomical analysis of sequencing reads revealed no sequence similarity to any known viruses. De novo assembly tools were then applied to the sequencing reads to construct contigs, three of which had some similarities to the L, M, and S segments of viruses from the Orthobunyavirus genus. High-quality, full-length genomic sequences of the three genomic segments (L, M, and S) of a novel Orthobunyavirus were produced after further refinement of these contigs. Further validation was done through the characterization of the genomic sequence, which included the prediction of open reading frames, examination of consensus genomic termini, and phylogenetic analysis. The new species of virus was then named the Ness Ziona virus.

10.3 NUCLEIC ACID-BASED DETECTION METHODS

Nucleic acid-based detection relies heavily on PCR, which amplifies DNA strands by undergoing numerous stepwise temperature cycles and a polymerase (Mullis et al. 1992). Real-time quantitative PCR (real-time qPCR) measures the target amplicon's synthesis throughout the experiment. This is made possible by DNA-intercalating dyes like SYBR green or fluorescently tagged probes. DNA dyes will adhere to all double-stranded DNA (Figure 10.1). The primers need to be highly optimized and should not produce non-specific amplicons as these non-specific products will also produce signals, distorting quantification. Fluorescently labeled probes are a more target-specific alternative (Burbelo et al. 2019). Numerous probe types are used in qPCR, but popular probes include those that must attach to a certain area of the target DNA in order for fluorescence to be obtained (e.g., hydrolysis or hybridization probes) (Figure 10.2).

Real-time qPCR has advantages and disadvantages with regards to the development of quick virus diagnostic tests. The advantage of utilizing probes like hydrolysis or hybridization probes is high specificity as both the primers and the probe must bind the target region to achieve signal. When there is an urgent need for a quick analysis, it is highly beneficial to use

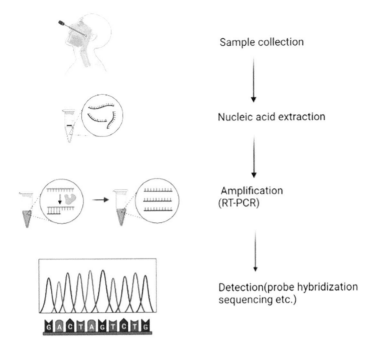

Figure 10.1 Nucleic acid detection process.

an analysis that yields highly precise, quantitative results. The turnaround time for findings from qPCR is reasonably short in terms of the actual test, but sample preparation, which is typically necessary for qPCR, can cause delays. While there are numerous commercially available kits that enable quick isolation of DNA or RNA, they have the drawback of being more expensive (Cassedy et al. 2021). This could prevent developing regions from using detection technologies like qPCR due to the upfront expense of the equipment itself. The ability of a quick assay to do high-throughput testing is another important feature.

Generally, qPCR is used in a 96-well format, thus the transition to high throughput is achievable. This is especially relevant in cases where high-volume testing is essential, such as when pandemic or epidemic samples need to be analyzed simultaneously, or in industrial or agricultural settings. This high-throughput capability of qPCR is complemented by its ability to be done in multiplexed reactions. Testing for virus panels is made possible by using multiple pairs of targeted probes and specialized primers that may be detected in the same sample but target various viruses. This provides a strong technique to simultaneously screen many samples for the presence of various viruses, especially when combined with established high-throughput technologies and the multi-well architecture of qPCR assays (Burbelo et al. 2019).

Fluorescent Dye-Based Real Time PCR (qPCR)

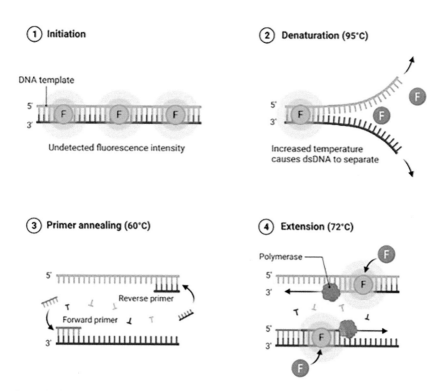

① **Initiation**

DNA template

5'
3'

Undetected fluorescence intensity

② **Denaturation (95°C)**

5'
3'

Increased temperature
causes dsDNA to separate

③ **Primer annealing (60°C)**

5'

Reverse primer

Forward primer

3'

④ **Extension (72°C)**

Polymerase

5'
3'

5'
3'

Figure 10.2 Fluorescent-based virus detection.

10.4 IMMUNOASSAY-BASED VIRAL DIAGNOSTICS

Antibodies are the primary tool used in immunoassays to identify viruses in a sample. There are three types of antibodies: polyclonal, monoclonal, and recombinant. Polyclonal antibodies are polyspecific because they bind to a variety of distinct structures or epitopes on the target molecule or organism (Magalhaes et al. 2022). This can be highly beneficial when the aim of the study is to extract all viral strains from a sample or when the desired result is the indiscriminate detection of all viral strains. Contrarily, monoclonal antibodies and some recombinant antibody variants exhibit single-epitope specificity, making them monospecific (Cassedy et al. 2021). Since they allow for the focused detection of specific locations on the target molecule, specific antibodies are valued in diagnostics. This technique is helpful for differentiating between several isolates or genotypes of a virus in virus diagnosis.

It takes time to find a reliable disease-specific antigen, isolate the best antibodies, and confirm the resulting antibodies and tests during the initial development of pathogen-targeting antibodies. However, once created, the antibody proteins are produced quickly. Additionally, the development of recombinant antibody technology, which produces antibodies as fragments of the entire antibody protein, has sped up and reduced the cost of producing antibodies (Magalhaes et al. 2022). The availability of cheaper expression hosts like yeasts and bacteria over mammalian cell culture reduces the cost of production of antibodies (Cassedy et al. 2021). Overall, immunoassays in viral detection play an essential role. They do have the advantage of lower costs, less complexity, and greater usability for usage by untrained persons, even though they may not always give the same sensitivity as nucleic acid detection methods, particularly at the early stages of infection.

10.5 LIPS PROFILING OF VIRAL ANTIBODIES

Luciferase immunoprecipitation systems (LIPS) use luciferase-tagged antigens in a highly quantitative immunoprecipitation experiment to diagnose viruses, track the effectiveness of antiviral medications, and classify viral infections according to their severity as presented in Figure 10.3 (Burbelo et al. 2019). LIPS have also been used to identify and find new viruses because the presence of host antibodies is proof that a virus is an infectious agent. LIPS were employed in one study to formally classify the astrovirus HMOAstV-C as a common novel

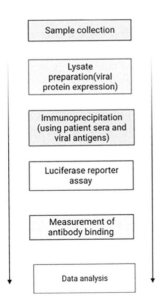

Figure 10.3 LIPS profiling process.

infectious agent in both children and adults (Burbelo et al. 2011). Noticeably, further independent molecular DNA sequencing tests of viruses discovered in cerebral spinal fluid from patients with unexplained brain inflammation have revealed that HMOAstV-C is the causing virus in some cases of viral encephalitis. These results highlight the value of developing new methods to categorize all viral infections and the potential influence that this knowledge may have on the identification of as-yet unidentified causes of human disease. New animal viruses have also been found and described using LIPS in addition to human viruses (Burbelo et al. 2019).

10.6 RNAscope FOR VIRAL PATHOGEN DETECTION AND VISUALIZATION

By using RNAscope in situ hybridization (ISH), which directly detects viral RNA in human or other animal cells, it is possible to determine the etiology and pathophysiology of viral illnesses. Viral detection has been revolutionized by nucleic acid-based molecular detection techniques, which provide several crucial benefits like sensitivity, specificity, and speed. In addition to the previously mentioned benefits, RNAscope ISH delivers molecular detection in conjunction with morphological context, which enables visualization of the virus in various infected tissues and cell types. Regarding sensitivity, RNAscope ISH consistently detects individual viral particles in infected cells despite low or undetectable viral loads with single RNA molecule detection technology.

Due to its high sensitivity, infections can be found even in their early stages. This technique is highly specific based on a proprietary probe design method that provides accurate detection even in the presence of contaminants or viruses that are identical to or related to the target viral species/strains. The rapid design and fabrication of new probes in single-day workflow coupled with 2 weeks makes RNAscope ISH a responsive method for finding novel or unusual viruses. It is possible to find numerous viruses or co-infection using RNAscope duplex and multiplex assays. Also, the flexible design of the probe targeting the sense or anti-sense strand enables the differentiation and detection of virus during latent or active stages (Atout et al. 2022).

10.7 MASS SPECTROMETRY–BASED DETECTION AND IDENTIFICATION OF RNA VIRUSES

Mass spectrometry (MS) based on proteomics is emerging as an efficient technique used as a diagnostic tool for various infectious diseases as shown in Figure 10.4 (Milewska et al. 2020). Recently, the MS-based approach is used for the diagnosis of rabies virus (RABV). Human rabies is an encephalitic illness which is transmitted by animals infected with lyssaviruses. In certain cases, the RABV incubation time in humans can last from a few weeks to several months. Neither antibodies nor viruses are found at this prodromal stage. At the stage

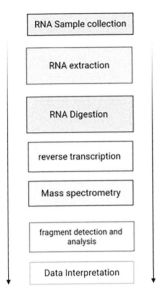

Figure 10.4 Mass spectrometry-based detection process.

of start of encephalitic symptoms, when the prognosis is virtually 100% deadly, antibodies, antigens, and nucleic acids are detected. Therefore, testing animals suspected of having rabies is the main intervention for human RABV exposure and subsequent post-exposure prophylaxis. Animal diagnostic procedures that are most frequently applied focus on RABV-encoded nucleoprotein (N protein) in the brain tissues as an antigen. N protein produced in the cytoplasm of infected cells can be seen as massive, granular inclusions in infected brain regions and can be detected by immuno-microscopy using anti-N protein. Reed et al. (2018) reported the detection of N protein by a novel MS-based method that is label-free and does not require target amplification. The MS-based method specifically detected N protein in brain tissue and identified RABV variants based on amino acid sequence information.

10.8 CONCLUSION

In this chapter, we discuss the applications of various innovative viral diagnostic methods. These quick viral diagnostic tests will probably have a significant effect on public health and affect how we monitor and manage various viral outbreaks. Currently, there is no such single method that could satisfy all the requirements for the viral diagnosis. There are some advantages and limitations associated with every diagnostic method. The ultimate requirement is the development of a technique which has the potential to provide low-cost diagnostics and portable tools to simultaneously detect many different viral agents in various clinical samples. Various standardized strategies are needed for the further implementation of these techniques (HTS, NGS, LIPS, RNAscope, MS, and others) in the clinical management of viral diagnostics.

REFERENCES

Atout S, Shurrab S, Loveridge C. Evaluation of the suitability of RNAscope as a technique to measure gene expression in clinical diagnostics: A systematic review. *Molecular Diagnosis & Therapy* 2022; 1–9.

Burbelo PD, Ching KH, Esper F, Iadarola MJ, Delwart E, Lipkin WI, Kapoor A. Serological studies confirm the novel astrovirus HMOAstV-C as a highly prevalent human infectious agent. *PloS One* 2011;6(8); e22576.

Burbelo PD, Iadarola MJ, Chaturvedi A. Emerging technologies for the detection of viral infections. *Future Virology* 2019;14(1); 39–49.

Cassedy A, Parle-McDermott A, O'Kennedy R. Virus detection: A review of the current and emerging molecular and immunological methods. *Frontiers in Molecular Biosciences* 2021; 76.

Fullard JF, Lee HC, Voloudakis G, Suo S, Javidfar B, Shao Z, Peter C, Zhang W, Jiang S, Corvelo A, Wargnier H. Single-nucleus transcriptome analysis of human brain immune response in patients with severe COVID-19. *Genome Medicine* 2021;13; 1–3.

Hilton SK, Castro-Nallar E, Pérez-Losada M, Toma I, McCaffrey TA, Hoffman EP, Siegel MO, Simon GL, Johnson WE, Crandall KA. Metataxonomic and metagenomic approaches vs. Culture-based techniques for clinical pathology. *Frontiers in Microbiology* 2016;7; 484.

Magalhaes IC, Souza PF, Marques LE, Girão NM, Araújo FM, Guedes MI. New insights into the recombinant proteins and monoclonal antibodies employed to immunodiagnosis and control of Zika virus infection: A review. *International Journal of Biological Macromolecules* 2022;200; 139–150. doi: 10.1016/j.ijbiomac.2021.12.196. Epub 2022, Jan 5.

Milewska A, Ner-Kluza J, Dabrowska A, Bodzon-Kulakowska A, Pyrc K, Suder P. Mass spectrometry in virological sciences. *Mass Spectrometry Reviews* 2020;39(5–6); 499–522.

Mullis K, Faloona F, Scharf S, Saiki R, Horn G, Erlich H. Specific enzymatic amplification of DNA in vitro: The polymerase chain reaction. *Biotechnology Series* 1992; 17.

Pachetti M, Marini B, Benedetti F, Giudici F, Mauro E, Storici P, Masciovecchio C, Angeletti S, Ciccozzi M, Gallo RC, Zella D. Emerging SARS-CoV-2 mutation hot spots include a novel RNA-dependent-RNA polymerase variant. *Journal of Translational Medicine* 2020;18; 1–9.

Reed M, Stuchlik O, Carson WC, Orciari L, Yager PA, Olson V, Li Y, Wu X, Pohl J, Satheshkumar PS. Novel mass spectrometry-based detection and identification of variants of rabies virus nucleoprotein in infected brain tissues. *PLoS Neglected Tropical Diseases*. 2018;12(12); e0006984.

Shifman O, Cohen-Gihon I, Beth-Din A, Zvi A, Laskar O, Paran N, Epstein E, Stein D, Dorozko M, Wolf D, Yitzhaki S. Identification and genetic characterization of a novel Orthobunyavirus species by a straightforward high-throughput sequencing-based approach. *Scientific Reports* 2019;9(1); 1–0.

Sime-Ngando T. Environmental bacteriophages: Viruses of microbes in aquatic ecosystems. *Frontiers in Microbiology* 2014;5; 355.

Su W, Liang D, Tan M. Nucleic acid-based detection for foodborne virus utilizing microfluidic systems. *Trends in Food Science & Technology* 2021;113; 97–109.

Swanson II PA, McGavern DB. Viral diseases of the central nervous system. *Current Opinion in Virology.* 2015;11; 44–54.

Trubl, G., Hyman, P., Roux, S., Abedon, S. T. (2020). Coming-of-age characterization of soil viruses: A user's guide to virus isolation, detection within metagenomes, and viromics. *Soil Systems* 2020;4(2); 23.

Zanella MC, Lenggenhager L, Schrenzel J, Cordey S, Kaiser L. High-throughput sequencing for the aetiologic identification of viral encephalitis, meningoencephalitis, and meningitis. A narrative review and clinical appraisal. *Clinical Microbiology and Infection* 2019;25(4); 422–430.

Zhao L, Rosario K, Breitbart M, Duffy S. Eukaryotic circular rep-encoding single-stranded DNA (CRESS DNA) viruses: Ubiquitous viruses with small genomes and a diverse host range. *Advances in Virus Research* 2019;103; 71–133.

Chapter 11

An Insight into the Symptomatic Treatment Modalities for COVID-19

Gifty Sawhney, Parveen Kumar, Suraj P. Parihar, and Mohit Sharma

11.1 INTRODUCTION

The outbreak of coronavirus disease 2019 (COVID-19) was reported from Wuhan, China, in late 2019 (1). COVID-19 is caused by a novel coronavirus, severe acute respiratory syndrome coronavirus 2 (SARS-CoV-2) (2), a member of a large family of viruses called β-coronavirus. Its genome is single-stranded RNA of about 30-kb nucleotides, which encodes for four structural proteins: membrane protein (M), spike protein (S), nucleoside protein (N), and an envelope protein (E) (3). As shown in Figure 11.1, among these four structural proteins the S protein is of special interest because these glycoprotein spikes offer the virus a crown-like morphology under an electron microscope (4).

SARS-CoV-2 is genetically closely related to SARS-CoV-1. Both viruses mediate entry into the host cell through the ACE2 receptor (5,6). This receptor is expressed by cardiopulmonary tissues but also monocytes and macrophages.

SARS-CoV-2 is contagious, spreading either by direct contact or through droplets spread by the coughing or sneezing of an infected person or from a contaminated surface. SARS-CoV-2 has an incubation period of 2.1 to 11.1 days, similar to that of Middle East respiratory syndrome (MERS) and SARS-CoV (7).

The median incubation period of SARS-CoV-2 is 5.1 days, and 97.5% of the symptomatic patients will develop symptoms within 11.5 days (8). Infected patients experience mild symptoms including cough, fever, fatigue, muscle pain, loss of appetite, and shortness of breath (9). About 15% of COVID-19 patients develop severe to critical complications that include acute respiratory distress syndrome, pneumonia, septic shock, and multiple organ failure and may also lead to death (9). As of May 2023, more than 765 million confirmed cases and more than 6.9 million deaths have been reported globally based on statistics from the World Health Organization (WHO). The number of infected patients is still increasing at an alarming rate. Lockdowns, massive screenings, quarantines, and social distancing have become the norm across the globe to prevent the spread of disease. Currently, no specific antiviral treatment capable of neutralizing the virus is available,

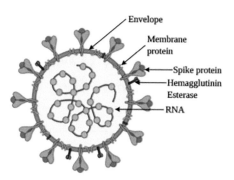

Figure 11.1 Structure of SARS-CoV-2.

hence the treatment given to COVID-19 patients is symptomatic and supportive. The mainstay of supportive therapy includes fluid management, oxygen therapy, and mechanical ventilation (10). A wide variety of drugs are repurposed for the treatment of COVID-19, including existing antiviral drugs: remdesivir, favipiravir, ribavirin, interferon, and lopinavir-ritonavir (an HIV protease inhibitor). Administration of antiviral drugs shortly after the onset of symptoms may reduce the chance of spreading infection by diminishing viral load in the secretions of the respiratory tracts of COVID-19 patients (11).

Antimalarial/anti-inflammatory drugs chloroquine (CQ), hydroxychloroquine (HCQ), tocilizumab, and corticosteroid are also administered in critically ill patients to reduce the chance of systemic inflammation before it results in multi-organ failure. Most of the studies suggest severe ill patients typically underwent deterioration in 1–2 weeks after onset, prompting the use of the anti-inflammatory drugs (10). Among the possible anti-COVID-19 treatments, passive antibody transfusion of convalescent plasma (CP) was also used for the emergency treatment of COVID-19 patients (12). The mechanism of action of CP includes a process of passive immunization (PI) through the transfer of neutralizing antibodies from recovered individuals to COVID-19 patients. Individuals who had recovered from COVID-19 (two consecutive negative polymerase chain reaction tests) at least 14 days earlier were eligible for plasma donation (13). However, there is no medical evidence to confirm the exact efficacy and safety of antiviral immunomodulatory drugs and CP in the treatment of COVID-19 patients.

11.2 METHODS UNDERTAKEN FOR ANALYSIS

We searched PubMed, LitCovid, and Google Scholar using the following terms: COVID-19, SARS-2, drugs, treatment, anti-inflammatory drug, antiviral, convalescent plasma, and vaccines. The search resulted in 200 articles,

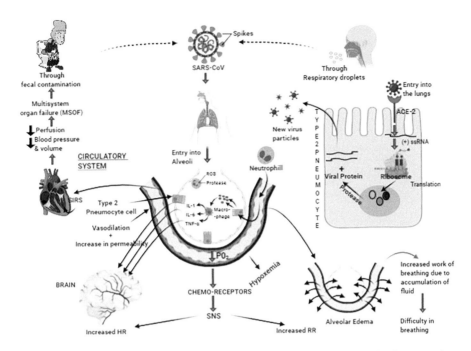

Figure 11.2 Mechanism of action of SARS-CoV. As the virus enters the body, it mainly alters the functioning of the circulatory system and respiratory system leading to increased respiratory rate and increased heart rate, ultimately leading to difficulty in breathing, decreased blood pressure, and decreased blood volume.

from among which relevant articles were selected. We also accessed reports from the WHO and Centers for Disease Control and Prevention. The authors reviewed the titles and abstracts for inclusion. The data was compiled from all the relevant scientific literature and an endeavor was made to review COVID-19 pandemic treatment and progress of vaccine development. As shown in Figure 11.2, intake of SARS-CoV can lead to various complications inside the human body affecting the digestive, circulatory, and respiratory systems ultimately leading to death.

11.3 THERAPEUTIC APPROACH AND OUTCOMES

The treatment option for chronic sequelae of COVID-19 is supportive. For the treatment of COVID-19, apart from supportive therapy, several treatments have been implemented, with varying success rates. In addition to the COVID-19 vaccination, several drugs have been identified based on their different mode of action on the virus and various pathways it navigates, including antimalarial drugs (HCQ/CQ); immunomodulatory agents (tocilizumab and corticosteroids); antiviral drugs (LPV/r, remdesivir, interferon,

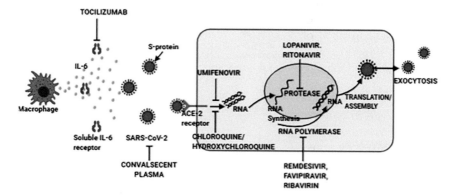

Figure 11.3 The target mechanism of action of different antimalarial, immunomodulatory, antiviral, anti-helminthic drugs available in the market.

favipiravir, and ribavirin), anti-helminthic (ivermectin) and CP transfusion are also recommended for the treatment of severe COVID-19 patients as presented in Figure 11.3 (14).

11.3.1 Remdesivir

Remdesivir is a nucleotide analogue manufactured by Gilead Science for the treatment of Ebola in 2016 (15). Remdesivir is incorporated in a growing RNA chain and evades proofreading by viral exoribonuclease and results in premature chain termination of viral RNA (16). This drug shows antiviral activities against different RNA viruses including SARS, MERS, and Ebola in in vitro and in non-human primate models. Owing to the promising antiviral activity of remdesivir against MERS and SARS-CoV-1, it was considered as a potential drug against SARS-CoV-2. Wang, Cao et al. conducted an experiment using Vero E6 cells to determine the anti–SARS-CoV-2 effect of remdesivir. It was found to be effective when administered 2 hours after infection at a multiplicity of infection of 0.05. However, no prophylactic effect was seen when administered prior to SARS-CoV-2 infection. The concentration for 90% of maximal effect (EC90) value of remdesivir against SARS-CoV-2 was found to be 1.76 µM. This study also suggested that remdesivir might inhibit SARS-CoV-2 in human hepatic cancerous Huh-7 cells (17). The current evidence on clinical and experimental studies indicated that remdesivir can be a good candidate for COVID-19 treatment. Non-human primates infected with MERS treated with a combined dose of interferon β and remdesivir were found to have improved respiratory function and decreased viral titer when compared to those treated with interferon β and LPV/r (18). In vivo efficacy of this drug was evaluated in rhesus macaques infected with SARS-CoV-2 showing striking effect with a reduction in viral titer and improvement in lung damage in treated animals (19). However, data from the compassionate use of remdesivir revealed that the treatment was associated with clinical improvement in 68% of the hospitalized patients with severe COVID-19 (16). A randomized, open-label, phase 3 trial found that hospitalized patients who did not require oxygen support and received a shorter, 5-day regimen experienced similar clinical improvement as patients who received a 10-day regimen (8). Holshue et al. reported that a COVID-19 patient in the United States was given remdesivir and showed improvement in clinical symptoms just after 2 days of treatment; the patient no longer needed oxygen support and his only symptoms were runny nose and dry cough (20). Many factors, including its intravenous route of administration and contraindication in renal impairment and those with liver enzyme levels elevated above five times the upper limit of normal range, limit the use of remdesivir. Nevertheless, for now, intravenous remdesivir has become a vital tool to improve outcomes for COVID-19 disease in hospitalized patients with pneumonia requiring supplemental oxygen treatment (21). As of October 22, 2020, the US Food and Drug Administration (FDA) has approved remdesivir (Veklury) for use in adult and pediatric patients of age 12 years and above and weighing at least 40 kg for the treatment of COVID-19 requiring hospitalization. Remdesivir is the first drug for COVID-19 to receive FDA approval. This approval was supported by the encouraging data from clinical studies involving remdesivir.

Adaptive COVID-19 Treatment Trial-1, conducted by the US National Institute of Allergy and Infectious Diseases, was randomized, double blind, and placebo controlled, and included 1062 patients. The treated group had a median recovery time of 10 days, compared to 15 days for the placebo group. Overall, clinical improvement at day 15 was higher in the remdesivir-treated group compared to the placebo group. The remdesivir-treated group had a mortality rate of 6.7%, compared to 11.9% in the placebo group (22).

A randomized, open-label multicenter clinical trial was conducted on hospitalized patients with severe COVID-19 in which patients received remdesivir for 5 or 10 days. The result of the study showed that COVID-19 symptom improvement was similar in the two groups. There were no significant differences in the mortality rates or recovery rates between the two groups (https://ClinicalTrials.gov Identifier: NCT04323761).

11.3.2 Lopinavir/Ritonavir (Kaletra)

Lopinavir is a protease inhibitor developed for the treat of HIV. It is co-formulated with ritonavir (LPV/r) to enhance the half-life of lopinavir through inhibition of CYP450 (23). Reports from various in vitro studies have revealed that SARS-CoV and MERS might be inhibited by lopinavir, with half-maximal effective concentration (EC50) at 6.6–17.1 µM (Choy, Wong et al. 2020). LPV/r has been shown to be effective against MERS infection in animal models and SARS infection in humans. It appears to be a powerful inhibitor of the main protease that exists in SARS-CoV-1 and inhibits the viral replication cycle (24).

With the surge of SARS-CoV-2, LPV/r was considered as one of the potential therapeutic options. LPV/r showed antiviral activity against SARS-CoV-2 in Vero E6 cells with the estimated EC50 at 26.63 μM (23). Moreover, LPV/r possesses considerable inhibitory activity against SARS-CoV-2 in vitro at concentrations (7/1.75 μg/mL) equivalent to their steady-state plasma levels. Currently, LPV/r at a dosage of 400 mg/100 mg orally twice daily in combination with ribavirin or without ribavirin is recommended for the treatment and management of COVID-19 patients in China (23). In a randomized clinical trial conducted on 86 COVID-19 patients in which 34 of them randomly received lopinavir/ritonavir, 35 received Arbidol, and the remaining 17 got no antiviral drug (2), the average period of positive to negative conversion was similar among all the groups. There was no difference among all the groups in the rates of symptoms improvement like cough easing, antipyresis, and development of chest CT at days 7 or 14 (25). A randomized, controlled, open-label trial in China showed that a twice-daily 2-week regimen of LPV/r for hospitalized adult patients with severe COVID-19 was not associated with significant benefits beyond the standard care. No differences were seen in the viral load clearance, improvement in clinical manifestation, and mortality rate between the two groups (26). Another clinically randomized, controlled, open-label trial found that hospitalized patients who received LPV/r was not associated with reduction in mortality, duration of hospital stay, or risk of progressing to invasive mechanical ventilation or death. These findings do not support the use of LPV/r for treatment of patients admitted to hospital with COVID-19 (27).

11.3.3 Favipiravir

Favipiravir (T-705; 6-fluoro-3-hydroxy-2-pyrazinecarboxamide) is a purine nucleoside analog that acts as a competitive inhibitor of RNA-dependent RNA polymerase (RdRp) and is known to be active in vitro against oseltamivir-resistant influenza viruses (28). In vitro studies indicated that favipiravir and oseltamivir possess a synergistic effect for influenza A viruses (28). Favipiravir is phosphorylated by cellular enzymes to favipiravir-ribofuranosyl-5′-triphosphate which the viral RNA polymerase recognizes as purine nucleotide (29). It has shown efficacy against a broad spectrum of RNA viruses. An open-label control study was designed to assess the efficacy of favipiravir in COVID-19 patients, wherein a 1600-mg dose of favipiravir twice per day on the first 2 days and 600 mg/day from days 3 to 14. This investigation has shown that favipiravir had significantly faster viral clearance and a higher improvement rate in chest imaging of treated patients compared to a control group treated with LPV/r (30). A clinical randomized trial demonstrated that in contrast to Arbidol, favipiravir did not conspicuously increase clinical recuperation rate at day 7 but significantly improved the latency to relief from pyrexia and cough (31). Another randomized, open-label controlled study, it was found that there was no significant differences in inflammatory biomarkers or clinical outcomes in COVID-19 patients with moderate to severe pneumonia treated with favipiravir and nebulizer-mediated interferon-1b against HCQ (32).

11.3.4 Arbidol (Umifenovir)

Arbidol is a selective broad-spectrum antiviral drug licensed in Russia and China for the prophylaxis and treatment of influenza and other respiratory viruses. Arbidol mainly functions by blocking the penetration of virus into host cells (33). It also displays immunomodulatory activity by inducing interferon production and macrophage activation (34). The in vitro antiviral activity of Arbidol against SARS-CoV-1 was assessed in GMK-AH-1 cells (35). It also significantly inhibited SARS-CoV-2 infection in vitro at low concentration (4.11 μm), suggesting that it could be effective in the treatment of COVID-19 (36). In a retrospective cohort study, it was found that Arbidol sped up the procedure of viral clearance and decreased demand for oxygen therapy in hospitalized patients (37). In another retrospective study it was found that combination therapy of Arbidol and LPV/r were effective in faster clearance of virus and chest CT scan improvement compared to patients receiving LPV/r monotherapy (38).

11.3.5 Ribavirin

Ribavirin is a nucleoside analogue of guanine antiviral drug that has been approved for treating several infections of viral causes, including hepatitis C virus and respiratory syncytial virus and has been evaluated in patients with MERS and SARS (39,40). Molecular docking studies predicted ribavirin might bind to SARS-CoV-2 RdRp but its in vitro activity has not been confirmed (41). The in vitro antiviral activity of ribavirin against SARS-CoV was estimated to be at a concentration of 50 μg/mL (39). Studies in mice revealed that ribavirin given during the first 3 days of infection increased SARS-CoV-1 lung viral titer and prolonged the duration of infection (42). In vivo study revealed serum concentration of ribavirin required to halt viral replication efficiently is higher than what is safely attainable in humans (43). In addition to the apparent lack of efficacy in most human studies, the risk of ribavirin-induced anemia along with hypoxia result in an increased risk of death in the treated SARS-CoV-1 patients (44). A clinically randomized, open-label trial was conducted to evaluate the efficacy and safety of ribavirin combined with LPV/r and IFN-β-1b in COVID-19 hospitalized patients (45). The antiviral combination significantly reduced the time to negative conversion by an average of 5 days and time to complete alleviation of symptoms by 4 days compared to the LPV/r control group. The recommended method of ribavirin for COVID-19 treatment was 500 mg intravenous infusion, two to three times per day, in combination with IFN-α or LPV/r, for 10 days maximum (46). However, its undesirable effect of reducing hemoglobin is harmful for patients in respiratory distress, and whether it offers significant potency against COVID-19 is uncertain.

11.3.6 Chloroquine and Hydroxychloroquine

CQ and its chemical cousin HCQ have been approved by the FDA for decades for the treatment of malaria and autoimmune diseases like rheumatoid arthritis (RA) and systemic lupus erythematosus (SLE). The antimalarial drugs (CQ and HCQ) have a broad spectrum of antiviral effect against RNA viruses (e.g., dengue, HIV, SARS-CoV, and MERS). Moreover, CQ and HCQ have shown immunomodulatory effects by suppressing the production and release of interleukin-6 (IL-6) and tumor necrosis factor alpha (TNF-α) (47,48). CQ and HCQ have shown to increase pH of endosome and interfere with the glycosylation of proteins. Wang et al. demonstrated the in vitro activity of HCQ against SARS-CoV-2 with an effective concentration EC50 at 48 hours of 1.13 μM in Vero E6 cells. It affects the glycosylation process of ACE2 receptors, thus causing the Vero cells treated with CQ to be refractory to SARS-CoV infection that may be the mechanism through which human cell can become refractory to COVID-19 infection. The state council of China, in a news briefing on February 17, 2020, said CQ showed acceptable safety and efficacy in treating COVID-19 (49). It has been reported that results from more than 100 COVID-19 patients demonstrated that CQ was effective in inhibiting the exacerbation of pneumonia, improving lung imaging findings, virus-negative conversion, and reducing the disease period (49). HCQ has more tolerability than CQ and has lower potential for a drug-drug interaction than CQ. It prevents the maturation of lysosome in antigen-presenting cells (dendritic cell and B cell), inhibits the processing of major histocompatibility complex 2–mediated antigen presentation to T cell. This will reduce T cell activation, differentiation, and proinflammatory cytokine production (IL-6, TNF-α, and IL-1) (50).

In addition to serving as anti-inflammatory agent, HCQ and CQ could inhibit the process of receptor binding and membrane fusion, two important steps needed for the host cell entry by coronaviruses. Gautret, Lagier et al. published their initial experience on the effect of 200 mg of HCQ on COVID-19 patients. The authors reported on 36 COVID-19 patients (20 HCQ and 16 controls). The investigators demonstrated that HCQ was superior to the standard of care in eliminating SARS CoV-2 from nasopharynx. In addition to HCQ, six patients were given azithromycin to prevent bacterial superinfection, and the investigators found that viral clearance was superior in these patients compared to those who received only HCQ (51). Their report also suggests that the effect of HCQ was higher in symptomatic patients than asymptomatic. The FDA in the latest treatment guidelines does not include the emergency use authorization (EUA) of CQ and HCQ in COVID-19 patients (Letter Revoking EUA, 2020). FDA has concluded that CQ and HCQ might not be effective in COVID-19 treatment and the benefits of these two drugs outweigh their known potential risk.

On June 17, 2020, the WHO stopped the HCQ arm of solidarity trial undertaken to find an effective treatment.

11.3.7 Tocilizumab

The elevated plasma level of IL-6 in SARS-CoV-2-infected patients is accompanied by an increase in C-reactive protein levels and lymphopenia, which together indicate a severe infection of SARS CoV-2. Tocilizumab a humanized monoclonal antibody, the antagonist of proinflammatory cytokine IL-6 receptor (52). It binds to soluble and membrane-bound IL-6 receptor and abrogate classical and trans-signaling cascade, rendering IL-6 incapable of immune damaging the cells and ameliorate the inflammatory response (53). Tocilizumab is used in the treatment of RA and cytokine releasing syndrome (CRS) (54). In cytokine release syndrome, this molecule neutralizes the main inflammatory factor and may lead to a block of cytokine storm during the systemic hyperinflammation stage (55). In COVID-19 patients, an elevated level of inflammatory cytokines was observed, including IL-6, due to the high concentration of inflammatory cytokines in blood serum vascular permeability enhances, resulting in accumulation of fluid and blood cells in alveoli of lungs, resulting in respiratory impairment. Therefore disruption in the IL-6 signaling pathway may be a possible treatment for COVID-19 infections. Data from the clinical study of Xu, Han et al. from China showed that the patients treated with tocilizumab showed relief from clinical symptoms: the body temperature returned to normal, and peripheral oxygen saturation improved and remained stable. Apart from this lymphocyte percentage in 52.6% patients and C-reactive protein in 84.2% of patients returned to normal on the fifth day after treatment (56). A randomized, double-blind placebo trial was conducted on severe COVID-19 patients with hyperinflammatory states, fever, and pulmonary infiltrates or need of oxygen therapy (57). Results of this trial showed that tocilizumab was not efficient for preventing intubation or death in moderately sick COVID-19 patients. Patients who got tocilizumab had fewer serious infections than patients who received placebo. However, further preclinical and clinical investigations are required to confirm its safety and efficacy.

11.3.8 Corticosteroids

Corticosteroids are immune-modulatory and anti-inflammatory drugs now employed for the treatment of COVID-19 (58). During the MERS and SARS-CoV outbreaks, corticosteroids were widely used, and now are being used for the treatment of COVID-19 along with other therapeutics (59). Various clinical studies revealed the efficacy of corticosteroids in the treatment of coronaviruses, but no consensus has been reached. Methylprednisolone (1–2 mg/kg per day) plays a pivotal role in reducing systemic inflammation and exudative fluid in the lungs, and prevents alveolar damage which can alleviate hypoxemia and protects the lung from further respiratory insufficiency in COVID-19 patients (60). A proper dose of corticosteroids could improve fever, pneumonia, and better oxygenation (61). Chen, Qi et al. reported that methylprednisolone is suggested for patients with acute respiratory distress syndrome for a shorter duration. However, corticosteroid therapy may suppress the immune response and delay virus clearance.

According to international guidelines for the management of septic shock and sepsis, glucocorticoids should be used in small doses and for a short period in critical cases on whom vasopressor therapy and adequate fluids do not restore hemodynamic stability (58). Wang, Jiang et al. conducted a clinical study of methylprednisolone on 46 COVID-19 patients with pneumonia hospitalized at Wuhan Union Hospital in China. This study revealed that intravenous administration of methylprednisolone results in a reduction of the time interval for body temperature to come back to normal, faster improvement of oxygen saturation and absorption degree of lung focus (50). Because of their anti-inflammatory nature, corticosteroids are recommended to treat severe acute respiratory infection of viral nature (50). Dexamethasone is a glucocorticoid that has been used for the treatment of allergies, asthma, and RA (29) as an immunomodulator and anti-inflammatory drug. Dexamethasone has emerged with strong evidence of improving patients infected with SARS-CoV-2 and mortality in severely ill patients. The ongoing clinical studies regarding the use of dexamethasone in COVID-19 patients shows striking effect. The clinical results of the recovery among COVID-19 patients involving the use of dexamethasone at a low to moderate dose of 6 mg once daily for 10 days showed reduced mortality in patients on ventilator (62,63). The use of dexamethasone in patients who were receiving oxygen therapy but were devoid of ventilation support shows improvement and reduced the mortality by 20%. Patients in mild condition requiring no oxygen support shows no effect (62,64). In this context, the WHO COVID-19 treatment panel issued guidelines on September 2, 2020, and strongly recommended the use of dexamethasone 6 mg daily or 50 mg of hydrocortisone every 8 hours for 7 to 10 days in the treatment of severe and critical patients (65). The panel also issued conditional recommendation against using corticosteroids for the treatment of mild and moderate COVID-19 patients.

11.3.9 Ivermectin

Ivermectin, an anti-helminthic agent which has been in use since 1970, stimulates gamma aminobutyric acid–gated chlorine channels, which in turn causes hyper-polarization results in paralysis of infected organism (66). An in silico study investigated that the strongest binding affinity of ivermectin against four key enzymes of SARS-CoV-2 (spike protein, 3Clpro, Plpro, and RdRp enzymes) suggest it as a potential inhibitor of the RdRp (67). It acts as an immunomodulator against the host organism by activating neutrophils, increase in the levels of C-reactive protein, and IL-6. It has also been reported in an in vitro study that ivermectin acts against flaviviruses like dengue fever, Japanese encephalitis, and tick-borne encephalitis virus (68). Like other RNA viruses, ivermectin also shows a positive role in eliminating coronaviruses by inhibiting importin α/β_1 mediated transport of viral protein in and out of the nucleus importins, a class of soluble transport receptors which help in nucleo-cytoplasmic transit of various substrates (69). Ivermectin produced by soil-dwelling actinomycetes, *Streptomyces avermitilis*, also shows inhibition of the chikungunya virus (CHIKV) in a baby hamster kidney BHK-21 cell line and a CHIKV replicon-containing cell line (BHK-CHIKV) in a dose-dependent manner. It has also shown antiviral activity against other alpha viruses like Semliki Forest virus, Sindbis virus, and yellow fever virus (70). Recently, in vitro studies using Vero-hSLAM cells has shown ivermectin inhibits the coronavirus (SARS-CoV-2) when a single dose is added to these cells (71). From these studies, it becomes imperative that this drug can provide beneficial effects to those patients who have already developed complications due to COVID-19 infection.

11.3.10 Interferons

Interferons are group of cytokines secreted by various cell types upon stimulation of pattern recognition receptor by viral components (72). Type 1 interferon is among the first cytokines produced during a viral infection (44). Type 1 interferon possesses a broad range of antiviral activities, exhibiting both direct inhibitory effect on viral replication and supporting an immune response to clear virus infection (73). It has been documented for decades that type-1 IFN-α can be used in the treatment of respiratory ailment of viral nature (65). IFN-α and IFN-β both show anti-SARS-CoV activity in vitro (74). IFN-β has also displayed potent activity against MERS in vitro with a half maximal inhibitory concentration of 1.37 U/mL. Some studies reported that IFN-β was superior against SARS-CoV than IFN-α (75). Moreover, IFN-α has synergistic effects when used with ribavirin, IFN-β, or IFN-γ (46).

From the data presented above, IFN-1 might be a safe and efficient treatment against SARS-CoV-2. Knowledge acquired during studies on SARS-CoV would be critical to assess in that perspective. In an in vitro study, SARS CoV-2–infected VERO E6 cell following type 1 IFN-α treatment showed significant reduction in viral replication (76). In China, the guidelines for the treatment of COVID-19 recommended 5 million units of IFN-α by vapor inhalation twice a day for a maximum of 10 days. It is necessary to use 2 mL of sterile water for injections as well (16). The WHO expert groups conducted a mega trial of four existing antiviral drugs: remdesivir, HCQ, lopinavir, and interferon regimens on hospitalized COVID-19 patients. It was found that these drugs manifest no mortality reduction or decrease in initiation of ventilation or duration of hospitalization (77,78).

Further clinical and preclinical trials are required to study the efficacy and safety of these drugs for the treatment of COVID-19.

11.3.11 Convalescent Plasma

PI for the treatment of human infectious diseases and artificially acquired passive immunity can be traced back to the 20th century. PI is a technique to achieve immediate to short period immunization against pathogens by the administration of pathogen-specific antibodies. Since its intervention, it has demonstrated to be lifesaving for many infected patients (79). In the early 20th century, CP was used to treat various kinds of viral infections like poliomyelitis, measles, mumps, and influenza (80). The SARS-CoV-2 virus outbreak has turned the spotlight onto the possible use of CP in the treatment of COVID-19, given the lack of specific prophylactic and therapeutic options. COVID-19 CP may be used for either treatment of disease or prophylaxis of infection (10). In a prophylactic purpose, the benefits of CP administration are that it can prevent infection in those who are at high risk, such as healthcare workers and individuals with underlying health conditions. It is obvious from the literature that CP from recovered patients can be used to patients as medication (39). Currently, CP from COVID-19 recovered patients is used in the treatment of acute COVID-19 patients. A clinical study involved ten COVID-19 patients; all ten patients were given 200 mL of CP with neutralizing antibody titers above 1:640 as an addition to the antiviral therapy and supportive care (81). The median time from the onset of illness to CP transfusion was 16.5 days. After CP transfusion the clinical symptoms were improved along with an increase of oxyhemoglobin saturation within 3 days, and several parameters improved as compared to pretransfusion, including an increase in lymphocyte counts, decrease in C-reactive protein, and the viral titer was unnoticeable after transfusion in seven patients (18). In a case series study by Ahn, Sohn et al. on two COVID-19 patients with CP, as assessed by X-ray the density of bilateral infiltration on chest improved, CRP and IL-6 level in the plasma returned to normal range, the fever subsided, and oxygen demand decreased (82).

In another case study by Shen Wang et al., five critically ill COVID-19 patients with acute respiratory distress syndrome, severe pneumonia, and viral titer remained high despite intervention of antiviral drugs and mechanical ventilation. All five patients received CP with the neutralizing antibody titer greater than 1:1000. After CP therapy, body temperature returned to normal range within 3 days, four patients' sequential organ failure assessment score decreased, viral load also decreased, and three out of five turned negative within 12 days after CP administration (83). A retrospective study found that hospitalized patients with prolonged positivity of SARS-CoV-2 received plasma therapy experienced improvement in lung imaging, shortened length of hospital stay, and rapid decrease in viral load (84). In a clinical study conducted on 333 patients, in which 228 were assigned to receive CP and 105 to get placebo, the median time from the onset of symptoms to enrollment in the study was 8 days and hypoxia was the main severity criterion for enrollment. At day 30, no considerable difference was observed between the treated group and the placebo group in the clinical status and overall mortality (85). In a clinical study, it was found that male COVID-19 patients were likely to produce more high-titer anti-SARS-CoV-2 IgG antibodies than female (86). The peak level of anti-SARS-CoV-2 IgG antibody in CP is maintained for a short duration. Therefore, using plasma from recovered COVID-19 patients for the therapeutic intervention should be within 28 days after hospital discharge.

11.4 DISCUSSION

With respect to the distinctive exploitation of COVID-19 vaccines, numerous pharmaceutical agencies and establishments have additionally released tasks for vaccine development. Thanks to the global drive and campaigns that we have successfully achieved several novel vaccine technologies including mRNA vaccine Spikevax by Moderna, Covovax by Serum Institute of India, Vaxzevria by Oxford University-AstraZeneca, to name a few. Accordingly, in the near future, the manufacturing of a number of vaccines can also additionally bring about the challenge in scaling up production quickly due to the fact that infrastructure wanted will vary in distinctive feature of the vaccine type. Antibody-dependent enhancement problem is probably a concern regarding vaccine development. In addition to the sensible layout of novel therapeutics that concentrate on viral replication or immunopathology, presently rapid screening of therapeutic agents to repurpose FDA-authorized and nicely characterized agents are probably a more practicable method. Clinicians and researchers must integrate efforts to hastily evolve our belief of all aspects of COVID-virus infections and fill in the gaps regarding the emergence of this virus. Ultimately, we also can analyze from this pandemic that we want to enhance our ailment tracking and surveillance system to prevent this sort of severe disease spread. According to reports, certain positive drugs as mentioned in Table 11.1 are recognized to be effective in treating COVID patients (82). But the dearth of clinical data may render the scientific prognosis tough to predict. A good sense of skepticism is needed until more definitive data are available. Hastily recommending unapproved treatments has the potential to cause harm. The treatment for COVID-19 should be utilized under ethically approved, randomized, controlled trials if possible (87). Regarding the hastily changing treatment for COVID-19, we must consider the evidence and approach them with scientific scrutiny.

Table 11.1 Potential Repurposed Drug Candidates against COVID-19

Name of the Drug Candidate	Class	Route of Administration	Mechanism of Action	Side Effects	References
Remdesivir	Antiviral	Intravenous	Inhibits RNA-dependent RNA polymerase, prematurely blocking RNA transcription	Anemia, deranged liver function tests, renal injury, and hyperglycemia	(88)
LPV/r	Antiviral	Oral	Inhibition of papain-like protease and 3C-like protease	Acute kidney injury, secondary infections, and respiratory failure; gastrointestinal adverse events	(89)
Favipiravir	Antiviral	Oral	RNA polymerase inhibitor	Teratogenic effect, reduction in neutrophil count	(90)
Ribavirin	Antiviral	Oral	RdRp inhibitor	Reduces hemoglobin	(22)
Arbidol (umifenovir)	Antiviral and immunomodulatory	Oral	S protein/ACE2, membrane fusion inhibitor	Allergic reaction, increase in transaminases and gastrointestinal upset	(91)
CQ/HCQ	Antimalarial	Oral	Inhibits viral entry, increases endosomal pH; additional immunomodulatory effects through inhibition of cytokine production	Abdominal cramps, diarrhea, anorexia, vomiting, nausea, hypoglycemia, retinal toxicity, cardiovascular problems, neuropsychiatric and central nervous system effects	(92)
Tocilizumab	Immunomodulatory	Intravenous	IL-6 inhibitor	Thrombocytopenia, neutropenia, and hypercholesterolemia	(1)
Corticosteroids	Immunomodulatory	Intravenous	Suppression of pro-inflammatory cytokines	Short-term use: hyperglycemia Long-term use: hypertension, weight gain, and increased risk of infections	(33)

11.5 CONCLUSION

In summary, we need aid in confronting a loathsome infection with more terrific infectivity than the SARS-CoVpandemic of 2003. Despite advancements in the medical world, there is currently no validated and effective therapeutic measure to treat severe COVID-19 patients. Many drugs for the treatment of chronic COVID-19 symptoms such as brain fog, delirium, intense pain, intense fatigue, respiratory issues as well as psychotic manifestations are investigated for efficacy and safety against SARS-CoV-2. Remdesivir shows promising results in many clinical outcomes. Remdesivir is designed to obstruct the replication of the virus in the body and has become the first drug to get official approval from the FDA for the treatment of the disease. Lopinavir-ritonavir, favipiravir, and combination treatment with HCQ alongside azithromycin emerges to be an alternate option for COVID-19 therapy. For patients with SARS-CoV-2 infection, angiotensin-converting enzyme inhibitors also work. A low dose of corticosteroids might be suggested for the treatment of refractory shock in COVID-19 patients. Due to the variation in the physiological characteristics in populations, distinctive antiviral therapy should be performed to attain healthier clinical results and avoid adverse drug reactions. Some of the latest studies have found that these drugs are not very significant in controlling this viral disease, so the precautions are the only available tool that will help us to come out from this pandemic.

Abbreviations

ACE2	Angiotensin-converting enzyme-related carboxypeptidase
ARDS	Acute respiratory distress syndrome
COVID-19	Coronavirus disease 2019
CP	Convalescent plasma
CQ	Chloroquine
CT	Computed tomography
E	Envelope protein
EC50	Half-maximal effective concentration
EUA	Emergency use authorization
FDA	U.S. Food and Drug Administration
HCQ	Hydroxychloroquine
IFN	Interferon
IL-1	Interleukin-1
IL-6	Interleukin-6
LPV/r	Lopinavir/ritonavir
M	Membrane protein
MERS	Middle East respiratory syndrome
N	Nucleoside protein

PI	Passive immunization
RA	Rheumatoid arthritis
RdRp	RNA-dependent RNA polymerase
RSV	Respiratory syncytial virus
S	Spike protein
SARS-CoV-1	Severe acute respiratory syndrome coronavirus-1
SARS-CoV-2	Severe acute respiratory syndrome coronavirus-2
SLE	Systemic lupus erythematosus
TNF-α	Tumor necrosis factor alpha
WHO	World Health Organization

REFERENCES

1. Jordan RE, Adab P, Cheng K. Covid-19: Risk factors for severe disease and death. *British Medical Journal Publishing Group* 2020 Mar 26;368; m1198. doi: 10.1136/bmj.m1198. PMID: 32217618.
2. Gautret P, et al. Hydroxychloroquine and azithromycin as a treatment of COVID-19: results of an open-label non-randomized clinical trial. *International Journal of Antimicrobial Agents* 2020;56(1); 105949.
3. Wu Y, et al., Patients with prolonged positivity of SARS-CoV-2 RNA benefit from convalescent plasma therapy: A retrospective study. *Virologica Sinica* 2020;35(6); 768–775.
4. Ye M, et al. Treatment with convalescent plasma for COVID-19 patients in Wuhan, China. *Journal of Medical Virology* 2020 Oct;92(10); 1890–1901. doi: 10.1002/jmv.25882. Epub 2020 Jun 29. PMID: 32293713.
5. McKee DL, et al. Candidate drugs against SARS-CoV-2 and COVID-19. *Pharmacological Research* 2020; 104859.
6. Islam MT, et al. A perspective on emerging therapeutic interventions for COVID-19. *Frontiers in Public Health* 2020; 8.
7. Atzeni F, et al. The effect of drugs used in rheumatology for treating SARS-CoV2 infection. *Expert Opinion on Biological Therapy* 2020; 1–10.
8. Goldman JD, et al. Remdesivir for 5 or 10 days in patients with severe COVID-19. *New England Journal of Medicine* 2020 Nov 5;383(19); 1827–1837. doi: 10.1056/NEJMoa2015301. Epub 2020 May 27. PMID: 32459919.
9. Wehbe Z, et al. Molecular insights into SARS COV-2 interaction with cardiovascular disease: Role of RAAS and MAPK signaling. *Frontiers in Pharmacology*, 2020;11; 836.
10. Sarkar C, et al. Potential therapeutic options for COVID-19: Current status, challenges, and future perspectives. *Frontiers in Pharmacology* 2020;11; 1428.
11. Mitjà O, Clotet, B. Use of antiviral drugs to reduce COVID-19 transmission. *Lancet Global Health* 2020;8(5); e639–e640.
12. von Rhein C, et al. Comparison of potency assays to assess SARS-CoV-2 neutralizing antibody capacity in COVID-19 convalescent plasma. *Journal of Virological Methods* 2020; 114031.
13. Del Fante C, et al. A retrospective study assessing the characteristics of COVID-19 convalescent plasma donors and donations. *Transfusion* 2021 Mar;61(3); 830–838. doi: 10.1111/trf.16208. Epub 2020 Dec 14. PMID: 33231325.
14. Khuroo MS, Sofi AA, Khuroo M. Chloroquine and hydroxychloroquine in coronavirus disease 2019 (COVID-19). Facts, fiction & the hype. A critical appraisal. *International Journal of Antimicrobial Agents* 2020; 106101.
15. Kaddoura M, et al. COVID-19 therapeutic options under investigation. *Frontiers in Pharmacology* 2020; 11.
16. Grein J, et al. Compassionate use of remdesivir for patients with severe Covid-19. *New England Journal of Medicine* 2020;382(24); 2327–2336.
17. Wang M, et al. Remdesivir and chloroquine effectively inhibit the recently emerged novel coronavirus (2019-nCoV) in-vitro. *Cell Research* 2020;30(3); 269–271.

18. Sheahan TP, et al. Comparative therapeutic efficacy of remdesivir and combination lopinavir, ritonavir, and interferon beta against MERS-CoV. *Nature Communications* 2020;11(1); 1–14.

19. Williamson BN, et al. Clinical benefit of remdesivir in rhesus macaques infected with SARS-CoV-2. *Nature* 2020 Sep;585(7824); 273–276. doi: 10.1038/s41586-020-2423-5. Epub 2020 Jun 9. PMID: 32516797.

20. Holshue ML, et al. First case of 2019 novel coronavirus in the United States. *The New England Journal of Medicine* 2020;382(10); 929–936.

21. Richardson C, Bhagani S, Pollara G. Antiviral treatment for COVID-19: The evidence supporting remdesivir. *Clinical Medicine* 2020;20(6); e215.

22. Beigel JH, et al. Remdesivir for the treatment of Covid-19—Preliminary report. *New England Journal of Medicine* 2020 Nov 5;383(19); 1813–1826. doi: 10.1056/NEJMoa2007764. Epub 2020 Oct 8. PMID: 32445440.

23. Choy KT, et al. Remdesivir, lopinavir, emetine, and homoharringtonine inhibit SARS-CoV-2 replication in-vitro. *Antiviral Research* 2020;178; 104786.

24. Chu C, et al. Role of lopinavir/ritonavir in the treatment of SARS: Initial virological and clinical findings. *Thorax* 2004;59(3); 252–256.

25. Li Y, et al., Efficacy and safety of lopinavir/ritonavir or Arbidol in adult patients with mild/moderate COVID-19: An exploratory randomized controlled trial. *Med.* 2020;1(1); 105–113; e4.

26. Cao B, et al. A trial of lopinavir–ritonavir in adults hospitalized with severe Covid-19. *New England Journal of Medicine* 2020 May 7;382(19); 1787–1799. doi: 10.1056/NEJMoa2001282. Epub 2020 Mar 18. PMID: 32187464.

27. Horby PW, et al. Lopinavir–ritonavir in patients admitted to hospital with COVID-19 (RECOVERY): A randomised, controlled, open-label, platform trial. *Lancet* 2020 Oct 24;396(10259); 1345–1352. doi: 10.1016/S0140-6736(20)32013-4. Epub 2020 Oct 5. PMID: 33031764.

28. Wang Y, et al. Comparative effectiveness of combined favipiravir and oseltamivir therapy versus oseltamivir monotherapy in critically ill patients with influenza virus infection. *Journal of Infectious Diseases* 2020;221(10); 1688–1698.

29. Furuta Y, et al. Favipiravir (T-705), a novel viral RNA polymerase inhibitor. *Antiviral Research* 2013;100(2); 446–454.

30. Cai Q, et al. Experimental treatment with favipiravir for COVID-19: An open-label control study. *Engineering* 2020;6(10); 1192–1198.

31. Chen C, et al. Favipiravir versus Arbidol for clinical recovery rate in moderate and severe adult COVID-19 patients: A prospective, multicenter, open-label, randomized controlled clinical trial. *Frontiers in Pharmacology* 2021 Sep 2;12; 683296. doi: 10.3389/fphar.2021.683296. PMID: 34539392.

32. Khamis F, et al. Randomized controlled open label trial on the use of favipiravir combined with inhaled interferon beta-1b in hospitalized patients with moderate to severe COVID-19 pneumonia. *International Journal of Infectious Diseases* 2020;102; 538–543.

33. Blaising J, Polyak SJ, Pécheur EI. Arbidol as a broad-spectrum antiviral: An update. *Antiviral Research* 2014;107; 84–94.

34. Glushkov R, et al. Mechanisms of Arbidol's immunomodulating action. *Vestnik Rossiiskoi Akademii Meditsinskikh Nauk* 1999;3; 36.

35. Khamitov R, et al. Antiviral activity of Arbidol and its derivatives against the pathogen of severe acute respiratory syndrome in the cell cultures. *Voprosy Virusologii* 2008;53(4); 9–13.

36. Wang X, et al. The anti-influenza virus drug, Arbidol is an efficient inhibitor of SARS-CoV-2 in-vitro. *Cell Discovery* 2020;6(1); 1–5.

37. Xu K, et al. *Clinical Efficacy of Arbidol in Patients with 2019 Novel Coronavirus-Infected Pneumonia: A Retrospective Cohort Study*; 2020. https://papers.ssrn.com/sol3/papers.cfm?abstract_id=3542148

38. Deng L, et al. Arbidol combined with LPV/r versus LPV/r alone against Corona Virus Disease 2019: A retrospective cohort study. *Journal of Infection* 2020 Jul;81(1); e1–e5. doi: 10.1016/j.jinf.2020.03.002. Epub 2020 Mar 11. PMID: 32171872.

39. Jean SS, Lee PI, Hsueh PR. Treatment options for COVID-19: The reality and challenges. *Journal of Microbiology, Immunology and Infection* 2020;53(3); 436–443.

40. Li G, De Clercq E. Therapeutic options for the 2019 novel coronavirus (2019-nCoV). *Nature Reviews Drug Discovery* 2020;19(3); 149–150.

41. Elfiky AA. Ribavirin, remdesivir, sofosbuvir, galidesivir, and tenofovir against SARS-CoV-2 RNA dependent RNA polymerase (RdRp): A molecular docking study. *Life Sciences* 2020;253; 117592.

42. Kaddoura M, et al. COVID-19 therapeutic options under investigation. *Frontiers in Pharmacology* 2020;11; 1196.

43. Barlow A, et al. Review of emerging pharmacotherapy for the treatment of coronavirus disease 2019. *Pharmacotherapy: The Journal of Human Pharmacology and Drug Therapy* 2020;40(5); 416–437.

44. Barnard DL, et al. Enhancement of the infectivity of SARS-CoV in BALB/c mice by IMP dehydrogenase inhibitors, including ribavirin. *Antiviral Research* 2006;71(1); 53–63.

45. Hung IFN, et al. Triple combination of interferon beta-1b, lopinavir–ritonavir, and ribavirin in the treatment of patients admitted to hospital with COVID-19: An open-label, randomised, phase 2 trial. *Lancet* 2020;395(10238); 1695–1704.

46. Dong L, Hu S, Gao J. Discovering drugs to treat coronavirus disease 2019 (COVID-19). *Drug Discoveries & Therapeutics* 2020;14(1); 58–60.

47. Zhao M. Cytokine storm and immunomodulatory therapy in COVID-19: role of chloroquine and anti-IL-6 monoclonal antibodies. *International Journal of Antimicrobial Agents* 2020 Jun;55(6); 105982. doi: 10.1016/j.ijantimicag.2020.105982. Epub 2020 Apr 16. PMID: 32305588.

48. Zhou D, Dai SM, Tong Q. COVID-19: A recommendation to examine the effect of hydroxychloroquine in preventing infection and progression. *Journal of Antimicrobial Chemotherapy* 2020 Jul 1;75(7); 1667–1670. doi: 10.1093/jac/dkaa114

49. Gao J, Tian Z, Yang X. Breakthrough: Chloroquine phosphate has shown apparent efficacy in treatment of COVID-19 associated pneumonia in clinical studies. *Bioscience Trends* 2020 Mar 16;14(1); 72–73. doi: 10.5582/bst.2020.01047. Epub 2020.

50. Gies V, Bekaddour N, Dieudonné Y, Guffroy A, Frenger Q, Gros F, Rodero MP, Herbeuval JP, Korganow AS. Beyond anti-viral effects of chloroquine/hydroxychloroquine. Frontiers in Immunology 2020 Jul 2;11:1409. doi: 10.3389/fimmu.2020.01409.

51. Lin SH, et al. Coronavirus disease 2019 (COVID-19): Cytokine storms, hyper-inflammatory phenotypes, and acute respiratory distress syndrome. *Genes & Diseases* 2020 Dec;7(4); 520–527. doi: 10.1016/j.gendis.2020.06.009. Epub 2020 Jun 29. PMID: 32837983.

52. Morena V, et al. Off-label use of tocilizumab for the treatment of SARS-CoV-2 pneumonia in Milan, Italy. *European Journal of Internal Medicine* 2020 Jun;76; 36–42. doi: 10.1016/j.ejim.2020.05.011. Epub 2020 May 21. PMID: 32448770.

53. Zhang C, et al. The cytokine release syndrome (CRS) of severe COVID-19 and interleukin-6 receptor (IL-6R) antagonist tocilizumab may be the key to reduce the mortality. *international Journal of Antimicrobial Agents* 2020; 105954.

54. Guaraldi G, et al. Tocilizumab in patients with severe COVID-19: A retrospective cohort study. *Lancet Rheumatology* 2020;2(8); e474–e484.

55. Klopfenstein T, et al. Tocilizumab therapy reduced intensive care unit admissions and/or mortality in COVID-19 patients. *Medecine et Maladies Infectieuses* 2020;50(5); 397–400.

56. Chi Z, et al. *The Cytokine Release Syndrome (CRS) of Severe COVID-19 and Interleukin-6 Receptor (IL-6R) Antagonist Tocilizumab May Be the Key to Reduce the Mortality*, 2020. www. ncbi. nlm. nih. gov/pmc/articles/PMC7118634/pdf/main. pdf.

57. Stone JH, et al. Efficacy of tocilizumab in patients hospitalized with Covid-19. *New England Journal of Medicine* 2020;383(24); 2333–2344.

58. Chen C, et al., Thalidomide combined with low-dose short-term glucocorticoid in the treatment of critical coronavirus disease 2019. *Clinical and Translational Medicine* 2020;10(2).

59. Russell CD, Millar JE, Baillie JK. Clinical evidence does not support corticosteroid treatment for 2019-nCoV lung injury. *Lancet* 2020;395(10223); 473–475.

60. Zhang W, et al. The use of anti-inflammatory drugs in the treatment of people with severe coronavirus disease 2019 (COVID-19): The perspectives of clinical immunologists from China. *Clinical Immunology* 2020;214; 108393.

61. Wang Y, et al. Early, low-dose and short-term application of corticosteroid treatment in patients with severe COVID-19 pneumonia: Single-center experience from Wuhan, China. *MedRxiv* 2020, doi: https://doi.org/10.1101/2020.03.06.20032342.

62. Ledford H, Coronavirus breakthrough: Dexamethasone is first drug shown to save lives. *Nature* 2020;582(7813); 469–470.

63. Peter Horby JRE, Richard Haynes, Martin J. Landray, Wei Shen Lim, Marion Mafham, Edmund Juszczak, J. Kenneth Baillie, Thomas Jaki. Dexamethasone in hospitalized patients with Covid-19 — Preliminary report. *New England Journal of Medicine* 2021 Feb 25;384(8); 693–704. doi: 10.1056/NEJMoa2021436. Epub 2020 Jul 17. PMID: 32678530.

64. Li G, De Clercq E. Therapeutic options for the 2019 novel coronavirus (2019-nCoV). *Nature Reviews Drug Discovery* 2020;19(3); 149–150.

65. World Health Organization. Corticosteroids for COVID-19: Living guidance. *World Health Organization*, https://apps.who.int/iris/handle/10665/334125. License: CC BY-NC-SA 3.0 IGO, 2 September 2020.

66. Njoo FL, et al. C-reactive protein and interleukin-6 are elevated in Onchocerciasis patients after ivermectin treatment. *Journal of Infectious Diseases* 1994;170(3); 663–668.

67. Swargiary, A. Ivermectin as a promising RNA-dependent RNA polymerase inhibitor and a therapeutic drug against SARS-CoV2: Evidence from in silico studies, 2020, September 09, PREPRINT (Version 1). Available at Research Square, https://doi.org/10.21203/rs.3.rs-73308/v1

68. Mastrangelo E, et al. Ivermectin is a potent inhibitor of flavivirus replication specifically targeting NS3 helicase activity: New prospects for an old drug. *Journal of Antimicrobial Chemotherapy* 2012;67(8); 1884–1894.

69. Gupta D, Sahoo AK, Singh A. Ivermectin: potential candidate for the treatment of Covid 19. *Brazilian Journal of Infectious Diseases* 2020;24(4); 369–371.

70. Varghese FS, et al. Discovery of berberine, abamectin and ivermectin as antivirals against chikungunya and other alphaviruses. *Antiviral Research* 2016;126; 117–124.

71. Caly L, et al. The FDA-approved drug ivermectin inhibits the replication of SARS-CoV-2 in-vitro. *Antiviral Research* 2020;178; 104787.

72. Sallard E, et al. Type 1 interferons as a potential treatment against COVID-19. *Antiviral Research* 2020;178; 104791.

73. Zhou Q, et al., Interferon-α2b treatment for COVID-19. *Frontiers in Immunology* 2020;11; 1061.

74. Chibber P, et al. Advances in the possible treatment of COVID-19: A review. *European Journal of Pharmacology* 2020;883; 173372.

75. Stockman LJ, Bellamy R, Garner P. SARS: Systematic review of treatment effects. *PLoS Medicine* 2006;3(9); e343.

76. Lokugamage KG, et al. Type I interferon susceptibility distinguishes SARS-CoV-2 from SARS-CoV. *Journal of Virology* 2020;94(23); e01410–e01420.

77. Pan H, et al. Repurposed antiviral drugs for COVID-19; Interim who solidarity trial results. *New England Journal of Medicine* 2021 Feb 11;384(6); 497–511. doi: 10.1056/NEJMoa2023184. Epub 2020 Dec 2. PMID: 33264556.

78. Consortium WST. Repurposed antiviral drugs for COVID-19—Interim who solidarity trial results. *New England Journal of Medicine* 2021;384(6); 497–511.

79. Luczkowiak J, et al. Specific neutralizing response in plasma from convalescent patients of Ebola virus disease against the West Africa Makona variant of Ebola virus. *Virus Research* 2016;213; 224–229.

80. Casadevall A, Pirofski LA. The convalescent sera option for containing COVID-19. *Journal of Clinical Investigation* 2020;130(4); 1545–1548.

81. Duan K, et al. Effectiveness of convalescent plasma therapy in severe COVID-19 patients. *Proceedings of the National Academy of Sciences* 2020;117(17); 9490–9496.

82. Ahn JY, et al., Use of convalescent plasma therapy in two COVID-19 patients with acute respiratory distress syndrome in Korea. *Journal of Korean Medical Science* 2020;35(14).

83. Shen C, et al. Treatment of 5 critically ill patients with COVID-19 with convalescent plasma. *JAMA* 2020;323(16); 1582–1589.

84. Wu Y, et al. Patients with prolonged positivity of SARS-CoV-2 RNA benefit from convalescent plasma therapy: A retrospective study. *Virologica Sinica* 2020; 1–8.

85. Simonovich VA, et al. A Randomized trial of convalescent plasma in Covid-19 severe pneumonia. *New England Journal of Medicine* 2020.

86. Wu C, et al. Influential factor and trend of specific IgG antibody titer in coronavirus disease 2019 convalescents. Zhong nan da xue xue bao. Yi xue ban = Journal of Central South University. *Medical Sciences* 2020;45(10); 1172–1175.

87. Zuckerman S, Barlavie Y, Niv Y, Arad D, Lev S. Accessing unproven interventions in the COVID-19 pandemic: Discussion on the ethics of 'compassionate therapies' in times of catastrophic pandemics. *Journal of Medical Ethics* 2022 Dec;48(12); 1000–1005. doi: 10.1136/medethics-2020-106783.

88. von Rhein C, et al. Comparison of potency assays to assess SARS-CoV-2 neutralizing antibody capacity in COVID-19 convalescent plasma. *Journal of Virological Methods* 2021;288; 114031.

89. Owa AB, Owa OT. Lopinavir/ritonavir use in Covid-19 infection: Is it completely non-beneficial? *Journal of Microbiology, Immunology and Infection* 2020;53(5); 674–675.

90. Torequl Islam M, et al. A perspective on emerging therapeutic interventions for COVID-19. *Frontiers in Public Health* 2020;8; 281.

91. Sanders JM, et al. Pharmacologic treatments for coronavirus disease 2019 (COVID-19): A review. *JAMA* 2020;323(18); 1824–1836.

92. Khamis F, et al. Randomized controlled open label trial on the use of favipiravir combined with inhaled interferon beta-1b in hospitalized patients with moderate to severe COVID-19 pneumonia. *International Journal of Infectious Diseases* 2021;102; 538–543.

Chapter 12

Innovations in RNA Therapeutics and Sequencing Technology to Combat Future Viral Pandemics

Snehasmita Jena, Sudakshya Sucharita Lenka, Shaikh Sheeran Naser, Apoorv Kirti, Suresh K. Verma, and Mrutyunjay Suar

12.1 INTRODUCTION

RNA therapeutics are a new class of drugs that utilize the properties of RNA to treat diseases. The first use of an RNA therapeutic was in 1988 when doctors used it to treat cancer patients who had failed other therapies. Since then, there has been a significant increase in research on RNA therapeutics and their applications for treating many different conditions. RNA therapeutics are also being used to treat viral infections and potentially treat cancer and other chronic diseases that cannot be cured with traditional medicine. RNA therapeutics work by interfering with the function of specific genes that have been identified as being involved in disease processes.

RNA sequencing technology has been used to sequence the RNA molecules found in cells, allowing us to identify specific genes or proteins within those cells. RNA sequencing can also identify new genes or proteins that may be involved in disease processes or even indicate how a particular disease has developed over time. With the advent of next-generation sequencing (NGS) technology, we are closer than ever to being able to sequence our entire genomes to determine what genes are active or inactive in our bodies. This will give us a more accurate understanding of how our bodies respond to treatments once they start. The development of RNA therapeutics and NGS technology has the potential to revolutionize the way we combat future pandemics because it will allow us to identify pathogens faster than ever before, get a better understanding of how they spread through society, develop effective countermeasures against them, and more quickly respond when one does occur.

12.2 A BRIEF HISTORY OF RNA THERAPY

RNA therapeutics is a new approach to treating diseases. It differs from traditional therapies as it uses genetic material to target specific cells and deliver a drug payload directly into the cell. Our current generation of RNA therapies has advanced through numerous significant breakthroughs. In the 1960s, research on nucleic acids led to the discovery of mRNA (1). Chemically synthesized antisense RNAs can suppress the expression of a target RNA since RNA is a nucleic acid that may bind to other nucleic acids in a sequence-specific manner. Stephenson and Zamecnik published the first application of RNA base pairing in 1978. The antisense oligonucleotide targeting 35s RNA sequence of Rous sarcoma virus (RSV) inhibited the viral duplication and cellular transformation (2). Studies in the 1990s have shown that direct injection of mRNA into mouse skeletal muscle expresses corresponding proteins without any delivery systems (3). Following this, mice immunized with mRNA-encoding influenza nucleoprotein encapsulated liposomes developed anti-influenza cytotoxic T lymphocytes (CTL). It was the first mRNA vaccine to be tested (4). The US Food and Drug Administration (FDA) authorized the first antisense oligonucleotide drug for cytomegalovirus (CMV) retinitis. Fomivirsen can be used to help patients resistant to other CMV treatments (5). The discovery of RNA interference (RNAi) in the late 1990s marked the beginning of a new paradigm in system biology and novel drug application (6). The scientific community quickly recognized RNAi and used it widely in a relatively brief period. Initial studies on using synthetic small interfering RNA (siRNA) to disrupt the sequence of the hepatitis C virus (HCV) in adult mice in vivo quickly established the therapeutic potential of RNA interference. siRNAs suppressed the transgene expression in adult mice (7). The FDA approved the first RNA aptamer, pegaptanib, in 2004. It is an anti-vascular endothelial growth factor (VEGF) examined to treat neovascular age-related macular degeneration (8). In 2008, the first mRNA vaccination trial was performed in melanoma patients (9). The rapid technological advancement in RNAi led to the first clinical trials based on siRNA. In 2010, researchers systematically administered siRNA to melanoma patients via targeted nanoparticles. This research demonstrated that siRNA could specifically target human genes (10). Subsequent research following this trial led to the approval of the first siRNA-based drug. Patisiran was authorized in 2018 to treat hereditary transthyretin amyloidosis (11). Doudna and Charpentier received the Nobel Prize in 2020 for their work on CRISPR-Cas9 gene editing. The CRISPR-Cas9 system involves a piece of RNA with a guide sequence that attaches to a specific target sequence of DNA in a genome. This modified RNA recognizes the DNA sequence and binds to the Cas9 enzyme, which cuts the DNA at the targeted location.

12.3 CLASSIFICATION OF RNA THERAPEUTICS

RNA-based medications could target multiple stages in the expression of genes, both coding and non-coding. Antisense oligonucleotides (ASOs) can regulate splicing, and siRNAs can target mature mRNAs. Small and long

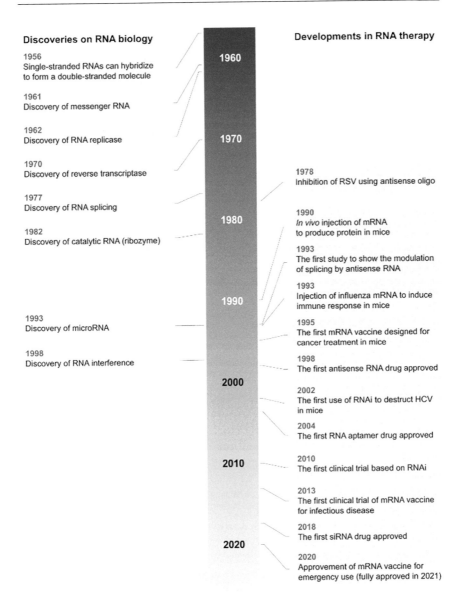

Discoveries on RNA biology

1956
Single-stranded RNAs can hybridize
to form a double-stranded molecule

1961
Discovery of messenger RNA

1962
Discovery of RNA replicase

1970
Discovery of reverse transcriptase

1977
Discovery of RNA splicing

1982
Discovery of catalytic RNA (ribozyme)

1993
Discovery of microRNA

1998
Discovery of RNA interference

Developments in RNA therapy

1978
Inhibition of RSV using antisense oligo

1990
In vivo injection of mRNA
to produce protein in mice

1993
The first study to show the modulation
of splicing by antisense RNA

1993
Injection of influenza mRNA to induce
immune response in mice

1995
The first mRNA vaccine designed for
cancer treatment in mice

1998
The first antisense RNA drug approved

2002
The first use of RNAi to destruct HCV
in mice

2004
The first RNA aptamer drug approved

2010
The first clinical trial based on RNAi

2013
The first clinical trial of mRNA vaccine
for infectious disease

2018
The first siRNA drug approved

2020
Approvement of mRNA vaccine for
emergency use (fully approved in 2021)

1960
1970
1980
1990
2000
2010
2020

Figure 12.1 The timeline of significant discoveries in RNA biology and advancements in
RNA therapy throughout history (12).

non-coding RNAs (ncRNAs) can be suppressed by ASOs or siRNAs.
Aptamer binding can alter protein function. Finally, exogenous mRNAs
can introduce specific proteins into cells to compensate for a deficient
enzyme or act as antigens to trigger a specific immune response.

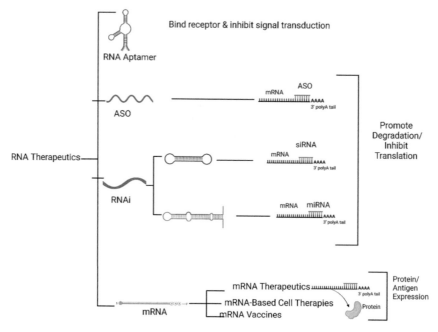

Figure 12.2 Various RNA therapeutics.

12.3.1 Antisense Oligonucleotides

ASOs are short, single-stranded oligonucleotides or RNA analogs that bind to target RNA regions by the conventional Watson-Crick base pairing. They can affect pre-mRNA splicing, trigger the breakdown of target mRNAs by RNase H-mediated degradation, or prevent the translation of mRNAs into proteins by binding to the target. ASOs are more stable and affine due to alterations to their backbone and sugar molecules (13). In the late 1970s, Zamecnik and Stephenson proposed using ASOs for therapeutic purposes. They developed tridecamer deoxynucleotides that matched the terminal sequences of Rous sarcoma virus (RSV) at both 3′ and 5′ ends. They then tested the effectiveness of these deoxynucleotides by introducing them to RSV-infected chick embryo fibroblasts, where they bonded with the complementary sequence in the target RNA (RSV RNA), preventing the production of the virus and subsequent cell transformation. The experiment successfully inhibited the virus and its effects (2).

ASOs can be classified into two categories based on their mechanism of action: The first category, Rnase H-dependent oligonucleotides, induces mRNA degradation (14). The second category are steric-blocker oligonucleotides, a type of molecule that can physically impede or hinder the process of splicing or the machinery involved in translation (15). The Rnase

H-dependent oligonucleotides are more commonly used for clinical purposes. Rnase H is a widely found enzyme that breaks down the RNA strand of an RNA/DNA hybrid. By using oligonucleotide, the reduction of targeted RNA expression through Rnase H can be highly effective, resulting in a down-regulation of protein and mRNA expression that can reach up to 80%–95% (16). In addition, they can be created in a manner that allows them to attach to specific locations on mRNA molecules within the cytosol, thereby hindering the translation process. An alternative method involves utilizing single-stranded ASOs to regulate the expression of microRNAs (miRNAs). These ASOs specifically attach to the target miRNAs, which results in the suppression of their function (known as antimiRs). Consequently, this reduces the activity of their related genes. ASOs can regulate pre-mRNA splicing in the specific gene by circumventing the mutation that causes the disease (17,18). Steric block ASOs function by physically hindering the process of translation or splicing. These ASOs can be designed to prevent polyadenylation, enhance or inhibit translation, or modify splicing (15,19–21).

Some ASO drugs with different chemistries and treatment targets have received FDA approval: nusinersen (Ionis Pharmaceuticals), eteplirsen (Sarepta Therapeutics), and inotersen (Ionis Pharmaceuticals and Akcea Therapeutics) (22–24). Efficient delivery of ASOs to their intended target site is a significant challenge when using them for therapeutic purposes. It is imperative that ASOs successfully reach the desired tissue and, subsequently, the appropriate intracellular compartment (25,26). Currently, the primary way to administer PS-modified single-stranded ASOs is through parenteral injection, including intravenous or subcutaneous injection, utilizing a basic

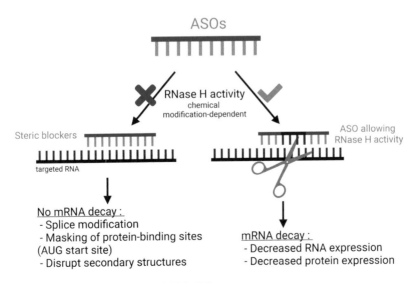

Figure 12.3 Mechanism of action of ASO (27).

saline solution (28,29). Despite the observation of ASO activity in several tissues, including the lung, stomach, bladder, and heart, it is noteworthy that the primary accumulation of ASOs occurs in the liver, kidney, bone marrow, adipocytes, and lymph nodes (29). Thus, to address delivery issues, it is vital to identify and analyze all obstacles that may hinder the movement of ASO within the body. This requires a thorough examination of all potential barriers. The approach should be comprehensive to ensure effective solutions are implemented. Two primary methods are being developed to enhance ASO delivery: viral and non-viral. Even though viral vectors are effective in delivering genetic material and can infect numerous cell types, they come with certain limitations, such as being prone to causing immune reactions, risking tumor formation, having limited loading capacity, and issues when it comes to scaling up (26).

12.3.2 RNA Interference

RNAi is a post-translational process wherein double-stranded RNA (dsRNA) is responsible for the targeted degradation of mRNA in a sequence-specific manner. RNAi is an intrinsic mechanism cells utilize to manage gene expression, which can hinder the translation of genes into proteins. Furthermore, cells also employ RNAi as a protective strategy against viral and bacterial nucleic acids, serving as an innate immune response (30). RNAi was initially discovered in plants (31). In 1998, Fire and Mello published a scientific paper that outlined a well-defined mechanism for how double-strand RNA (dsRNA) can effectively silence mRNA in a specific organism called *Caenorhabditis elegans* (6,32). The RNAi pathway involves two crucial small RNA molecules: siRNA and miRNA.

12.3.2.1 Small Interfering RNA

siRNAs are composed of non-coding RNA duplexes derived from precursor siRNAs. These precursor siRNAs are either produced through transcription or artificially introduced and can vary in size, ranging from 30 bp to over 100 bp. To initiate RNAi, the precursor siRNA duplex undergoes processing by the endogenous Dicer enzyme. This results in the formation of 20- to 30-bp-long siRNA with two base overhangs in the 3′ area, which interacts with the RNA-induced silencing complex (RISC). The RISC has the endonuclease argonaute 2 (AGO2), which cleaves the sense strand and leaves the antisense strand, guiding the active RISC toward the target mRNA. Once the active RISC locates the target mRNA, AGO2 cleaves its phosphodiester backbone. The antisense strand usually has complete complementarity with the coding region of the target mRNA, making siRNA effective in knocking down a specific target gene (33,34). While numerous drugs are undergoing clinical trials, the FDA has approved two siRNA drugs.

One of these is patisiran, a chemically altered siRNA medication utilized to manage hereditary transthyretin-mediated amyloidosis. The drug functions by binding to the 3′ untranslated section of transthyretin mRNA, resulting in its cleavage (11). In 2019, the FDA approved Givosiran, the second siRNA medication, for treating acute hepatic porphyria, an uncommon inherited genetic disorder. Its function is to attach to and inhibit the translation of delta-aminolevulinic acid synthase 1 (ALAS1) mRNA, reducing the neurotoxic intermediates present in this disease. This medication's approval is a significant development in the treatment of this particular ailment (35). When using siRNA-based therapeutic approaches, several challenges must be taken into account. These include problems with off-target effects, efficacy, delivery, and immune system activation (36,37). Transporting double-stranded siRNAs inside cells is more challenging than single-stranded ASOs (38). It is important to note that only a tiny quantity of cytosolic siRNAs per cell, likely in the hundreds, is required for gene knockdown to occur effectively and continuously (39,40). This is achieved due to the stability of the guide strand of the siRNA that remains constant within the RISC for weeks, despite being diluted with each cell division (41).

12.3.2.2 MicroRNAs

miRNAs are small RNA molecules that lack coding sequences and regulate the expression of several mRNAs. They achieve this by either impeding translation or stimulating the degeneration of the target mRNAs. It is widely believed that miRNAs manage the functioning of approximately 30% of human genes (42,43). The initial sighting of this set of non-coding RNAs was in *C. elegans*. These RNAs are produced as primary miRNAs (pri-miRNAs) by transcribing genomic DNA (44). The pri-miRNAs are transcribed from genomic DNA and belong to the non-coding RNA class. These pri-miRNAs have a loop structure with mismatches that Drosha cleaves to form precursor miRNAs (pre-miRNAs) that are 70–100 base pairs long.

Exportin 5 transports the pre-miRNAs to the cytoplasm, where Dicer processes them into RNA duplexes with two base overhangs in the 3′ region called miRNAs. The miRNAs are then loaded into the RISC to form a miRISC complex. The sense strand is released from the miRNA duplex, and the antisense strand guides the miRISC. The target mRNA is usually inhibited through translational repression, degradation, or cleavage by hybridization occurring at 2–7 bases of the 5′ end of miRNA and the 3′ UTR of the target mRNA (45–47). There are two types of miRNA-based therapeutics: miRNA mimics and miRNA inhibitors. miRNA mimics are double-stranded RNA molecules that imitate miRNAs, while miRNA inhibitors are single-stranded RNA oligos that obstruct miRNAs (40). Like siRNAs, miRNAs can also be tailored to target a specific gene. However, delivery, specificity, toxicity, and immune response challenges persist. Chemical modifications can optimize miRNAs to surmount these obstacles, but it is necessary to ensure that the miRNA can still incorporate into the RISC complex. Currently, targeting miRNAs with ASOs is more general and practical since miRNAs regulate many genes. ASOs targeting miRNAs have been proven safe and effective in mice, non-human primates, and humans (48–50).

12.3.3 RNA Aptamers

Oligonucleotide aptamers are a different category of promising RNA therapies. The term aptamer is derived from the Latin word *aptus*, which means "to fix." This choice of term is significant as it refers to the specific and complementary binding relationship of aptamers with their corresponding targets, like the concept of a lock-and-key mechanism (51,52). Aptamers are small oligonucleotides, typically between 20 and 70 bases, that are single stranded and can be made of either RNA or DNA. These molecules can bind to specific targets through a three-dimensional complementary structure, resulting in a high level of affinity and specificity. Unlike other methods, aptamers can be customized to target a range of substances, including nucleotides, amino acids, proteins, small molecules, and even live cells (53). In aptamer research, proteins remain the primary focus (54). Oligonucleotide aptamers possess similar affinity and specificity properties as monoclonal antibodies but with the additional benefits of minimal immunogenicity, high production, low cost, and high stability. The selection of these oligonucleotides can be achieved through a process called systematic evolution of ligands by exponential enrichment (SELEX), developed by Szostak and Gold in 1990 and involving an in vivo approach (51,53,54). From 2005 to about 2015, there was a significant increase in the popularity of oligonucleotide aptamers after the approval of the first aptamer, Pegaptanib, by the FDA for wet age-related macular degeneration therapy. These aptamers have been extensively used for diagnostic and therapeutic purposes against various targets, resulting in the development of over 900 aptamers until the last estimates were published online in 2016 (55). Despite the potential benefits, there is currently limited activity in the drug development of aptamers. However, these oligonucleotides can potentially replace therapeutic antibodies in specific applications with a lower risk of immune responses. Aptamers could also serve as a means for targeted intracellular delivery of RNA-based drugs and other molecules.

12.3.4 Messenger RNAs as Therapeutics

An additional technique involving RNA is the utilization of stabilized mRNAs that have undergone chemical modifications. These exogenous mRNAs will be translated into proteins over time. mRNAs are intermediates between the coding genomic DNA and encoded proteins (56). mRNAs act as the temporary blueprints for genes in the genomic DNA. The translational machinery then uses these instructions to produce the designated proteins. All genetic information can be included during this process (3,57). Recently, there has been an increased interest in using in vitro transcribed (IVT) mRNA as a new type of drug that can transmit genetic information. These synthetic mRNAs can be modified to temporarily express proteins and mimic natural mRNAs' structure (38,58). mRNA-based therapy offers a key benefit over viral gene delivery as mRNA does not pass into the nucleus, lowering the risk of insertional mutagenesis.

Additionally, mRNA enables transient protein expression dependent on the half-life, thus avoiding gene activation and maintaining dose responsiveness. This emerging therapy category, IVT mRNA treatment, is gaining popularity due to these advantages. Currently, several mRNA-based cancer immunotherapies and vaccines against infectious diseases are undergoing clinical trials (58,59). Despite the significant similarity between IVT mRNA and naturally occurring mRNA, it is noteworthy that the innate immune system can still identify it. This fact may have a significant impact on its applicability. In the context of vaccination approaches, the immune response triggered by mRNA-induced immune stimulation may be enhanced by the production of inflammatory cytokines, thereby improving its effectiveness. However, for non-immunotherapy treatments, the situation is distinct, and currently, mRNA-based therapeutics have only been tested in clinical trials for cancer immunotherapy (9,60–66).

The process of designing and producing RNA therapeutics is comparatively more straightforward and less expensive than that for recombinant proteins or small molecules. Furthermore, RNA sequences can be conveniently altered and personalized to meet the requirements of RNA therapy.

12.3.5 mRNA Vaccines

In the last 10 years, mRNA has emerged as a promising therapeutic tool for vaccine development and protein replacement therapy, thanks to significant technological advancements and research investment. mRNA has various advantages compared to subunit, killed and live attenuated viruses, and DNA-based vaccines. First, it is safe to use as it is a non-infectious and non-integrating platform, eliminating the risk of infection or insertional mutagenesis. Moreover, the cellular processes degrade mRNA, and its in vivo half-life can be regulated with specific modifications and delivery methods (67–69). Second, mRNA has been modified to enhance its efficacy by improving its stability and translatability. Third, mRNA vaccines have the advantage of being rapidly, inexpensively, and quickly produced on a large scale due to the high yields of in vitro transcription reactions. These modifications and production methods have made mRNA vaccines a promising option for combating diseases (67,70,71). mRNA-based vaccines have the potential to induce immunity against infectious diseases (prophylactic vaccines) and stimulate the immune system to combat cancer (therapeutic vaccines) by encoding antigens and adjuvants. RNA vaccines can trigger both cell-mediated and humoral immunity (59).

12.3.6 CRISPR-Cas Gene Editing

The scientific community has widely accepted a new genetic technology called CRISPR-Cas9, which stands for Clustered Regulatory Interspaced Short Palindromic Repeats and the CRISPR-associated protein 9. This technology is used for various genetic applications and has gained unprecedented popularity (72). Like RNAi, the CRISPR-Cas system was not created for gene editing. It aims to protect prokaryotes from foreign DNA invasions such as viruses and plasmids. This defense mechanism was naturally occurring and was discovered in bacteria and archaea (73–75). It was later modified for use in laboratory settings (75,76). The CRISPR system is divided into two categories, each with its variations. Class I CRISPR systems rely on various Cas proteins to break down foreign nucleic acids, while class II systems utilize one large Cas endonuclease (77). The CRISPR/CAS9 technology helps identify essential genes in various cellular pathways. It is more effective than earlier methods like retroviral insertional mutagenesis or RNAi libraries, with limitations such as incomplete inactivation or off-target effects. The CRISPR/Cas9 system enables the reliable identification of critical gene products through highly efficient and robust screening (78). The technology of CRISPR/Cas9 is an up-and-coming means of screening to understand cellular pathways better. This understanding can then be applied to analyze the factors that play a role in the manifestation and pathophysiology of infectious diseases, both in the host and the pathogen. The CRISPR/Cas9 system has been successfully reprogrammed to combat bacterial pathogens by utilizing its own mechanisms. This technique can assist in identifying the function of genes and potential targets for antibiotics in bacteria, which can be used to develop effective therapeutic intervention strategies. One of the most notable examples is *Mycobacterium tuberculosis*, a significant public health concern despite years of extensive research due to its high prevalence and mortality rate. With the rise of numerous strains of bacteria resistant to multiple drugs, it is crucial to find new ways to treat them. Chemotherapeutic agents and better vaccines are necessary solutions. An effective method of controlling the genes responsible for tuberculosis infection is through CRISPR interference (CRISPRi). This technique can repress target genes, which helps identify and study essential bacterial virulence genes and determine new targets for small molecule inhibition (79,80). The CRISPR/Cas9 system was first developed in bacteria to target foreign DNA and viruses, making it a potentially useful tool for treating viral infections. Researchers have successfully used CRISPR/Cas9 to modify the vaccinia virus, enhancing its ability to be a vector in vaccines for infectious diseases and cancer immunotherapies

Table 12.1 Summarizing Different RNA Therapeutics for Viral Diseases

RNA Therapy	Definition	Therapeutics	Target Virus	Status
ASOs	Short single-stranded oligonucleotides bind to target RNA regions, preventing their translation into proteins	AVI-6002 GSK2838232 Radavirsen (AVI-7100)	Ebola virus (EBOV) HIV H1N1 influenza virus	Phase I (81) Phase II (82) Phase I completed (83)
siRNAs	Small interfering RNA molecules target and degrade specific mRNA molecules, preventing the production of certain proteins	TKM-130803 MIR 19 (siR-7-EM/ KK-46)	EBOV SARS-COV 2	Phase II* (84) Phase II completed (85)
miRNAs	Small RNA molecules bind to complementary mRNA molecules, leading to the degradation of the mRNA or the inhibition of its translation into protein	Miravirsen RG-101	Hepatitis C virus (HCV) HCV	Phase II completed (50) Phase II* (86)
RNA aptamers	Small single-stranded RNA bind to specific targets through a three-dimensional complementary structure	ES15 39SGP1A Pseudoknot M302 RU25–80 DP6–22 R-F t1 P-58	SARS-COV 2 EBOV HIV HIV HIV HIV HCV HCV	Pre-clinical stage (87) Pre-clinical stage (88) Pre-clinical stage (89,90) Pre-clinical stage (91) Pre-clinical stage (92) Pre-clinical stage (93) Pre-clinical stage (94) Pre-clinical stage (95)

(Continued)

Table 12.1 Summarizing Different RNA Therapeutics for Viral Diseases (Continued)

RNA Therapy	Definition	Therapeutics	Target Virus	Status
mRNA	Deliver mRNA molecules to cells, which are then translated into proteins by the cell's own machinery	BNT162b2	SARS-CoV-2	FDA approved (96)
		BNT162b1	SARS-CoV-2	Phase II/III (97)
		mRNA-1273	SARS-CoV-2	FDA approved (98)
		CvnCoV	SARS-CoV-2	Phase III (99)
		ARCoV	SARS-CoV-2	Phase III (100)
		ARCT-021	SARS-CoV-2	Phase II (101)
		MRT5500	SARS-CoV-2	Phase I/II (102)
		CV7201	Rabies	Phase I completed (103)
		CV7202	Rabies	Phase I (104)
		AGS-004	HIV	Phase I (105)
		iHIVARNA-01	HIV	Phase III* (106)
		mRNA1325	Zika virus	Phase I (107)

*Withdrawn or terminated.

(108). A study demonstrated the feasibility of utilizing CRISPR/Cas9 as a tool for attenuating novel strains of viruses, specifically in the case of the live attenuated pseudorabies virus, which has potential implications for vaccine production (109). The CRISPR/Cas9 system, which has been developed as a genetic tool, has shown considerable potential in improving our ability to manage and eliminate several major infectious diseases worldwide in just a few years. Nevertheless, despite the advancements in comprehending the function of CRISPR, several essential aspects of the system still need to be clarified.

12.4 ADVANCEMENT IN RNA SEQUENCING TECHNOLOGY

All cells in the human body contain the same genetic material, but the expression of genes varies greatly, resulting in different characteristics of individual cells. In order to comprehend the function of genes in normal biological processes and diseases such as cancer, it is essential to analyze the transcriptome, which includes all transcribed RNA in an organism, such as mRNA, rRNA, tRNA, and non-coding RNA. RNA sequencing (RNA-seq) is a technique that employs NGS to identify the quantity and sequence of RNA in an organism, tissue, or individual cell. In the past, Sanger sequencing was used for RNA-seq, which was on the cutting edge at the time but limited in both speed and expense. However, with the introduction and widespread use of NGS technology, RNA-seq has reached its full potential (110). The process of RNA-seq involves many stages: RNA extraction, reverse transcription into cDNA, adapted ligation, amplification, and sequencing. Each of these steps is critical to the overall workflow of RNA-seq. To complete the process, specific computational tasks are required. These include aligning and piecing together the sequencing reads to form a transcriptome, determining the number of reads that overlap with transcripts, refining and standardizing the results across different samples and applying statistical analysis to identify substantial variations in the expression levels of specific genes and transcripts between different groups of samples. The advancement in technology in the wet lab and computational techniques has been the driving force behind RNA-seq. This method has provided a more comprehensive and unbiased understanding of RNA biology and transcriptome compared to previous microarray-based methods. So far, nearly 100 different approaches have been developed from the standard RNA-seq protocol (Illumina). The development of research methods has primarily focused on Illumina short-read sequencing tools. However, advancements in long-read RNA-seq and direct RNA sequencing, such as dRNA-seq, are now allowing users to explore inquiries that could not be addressed using Illumina short-read technologies (111–113).

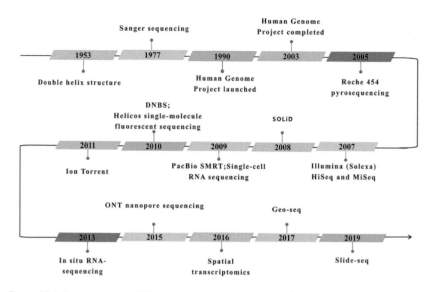

Figure 12.4 Development of RNA sequencing technologies (114).

12.4.1 Short-Read RNA-Seq

Short-read sequencing has become the commonly used approach for identifying and measuring gene expression throughout the transcriptome. This is mainly due to its cost-effectiveness and ease of use compared to microarrays. Short-read sequencing also provides detailed and reliable data that accurately captures quantitative expression levels throughout the transcriptome. Most publicly available RNA-seq data is produced using high-throughput sequencing-by-synthesis short-read technologies that can read fragments ranging from 50 to 500 base pairs. These technologies generate an average depth of 20–30 million reads per sample. The process involves extracting mRNA and generating cDNA, followed by adapter ligation and polymerase chain reaction (PCR) amplification to create clusters on a flow cell. Fluorescently labeled nucleotides are introduced during sequencing, and the complementary strand is imaged after each cycle. Computational methods are then used to identify individual transcripts and measure their abundance in differential gene expression studies. Although short-read RNA-seq provides high throughput and a wide range of protocols, it has inherent bias risks due to cDNA synthesis and PCR amplification steps. Furthermore, it is unsuitable for analyzing isoforms and long transcripts (115–117).

12.4.2 Long-Read RNA-Seq

Presently, Illumina sequencing is the most widely used platform for RNA-seq. However, long-read technologies offered by Pacific Biosciences (PacBio) and Oxford Nanopore (ONT) provide alternative options for single-molecule sequencing of complete RNA molecules after cDNA conversion. These technologies are gaining popularity as well (118–120). These methods eliminate the requirement of putting together brief RNA-seq reads, resolving some problems that arise with short-read techniques. This decreases ambiguity during sequence read mapping and helps identify longer transcripts, leading to a more comprehensive collection of isoform diversity.

Furthermore, these methods minimize the high occurrence of false-positive splice-junction detection often seen in short-read RNA-seq computational tools (121). Long-read technologies can sequence full-length transcripts at the single-molecule level, making them ideal for more complicated analyses such as the discovery of novel isoforms. The most prominent long-read technology is the single-molecule real-time (SMRT) platform, which converts RNA into cDNA through template-switching reverse transcriptase and PCR amplification to create enough templates. The sequencing process occurs on a chip, where individual cDNA molecules are bound to a single polymerase molecule fixed at the bottom of a nanowell. The well is imaged after each fluorescently labeled nucleotide is incorporated into the growing strand, and the sequence is then assembled and analyzed computationally (122). Although this technology has the benefit of identifying isoforms without requiring the assembly of transcripts from short reads, its throughput is significantly lower, ranging from 500,000 to 10 million reads per run. This limitation can hinder large-scale differential expression analysis studies. Moreover, there is a risk of introducing bias and errors in the reverse transcription and PCR amplification steps, which have error rates that are two orders of magnitude higher than the best short-read methods. Furthermore, unlike short-read methods, this protocol cannot be used for studies with degraded RNA.

12.4.3 Direct-Read RNA-Seq

The long-read techniques mentioned earlier, such as the primary short-read platform, necessitate the conversion of mRNA to cDNA before sequencing. However, Oxford Nanopore has recently proven that its nanopore sequencing technology can sequence RNA directly, without any changes or amplification during library preparation (111,113,123,124). This shows promising potential for more efficient sequencing methods. The dRNA-seq method eliminates biases caused by specific processes and keeps epigenetic information intact. RNA library preparation involves linking two adaptors in sequence. Initially, a duplex adaptor with an oligo(dT) overhang is attached and joined to the RNA polyadenylation [poly(A)] tail, followed by a reverse-transcription step that is optional but recommended to improve sequencing throughput. Then, the second ligation step attaches pre-loaded sequencing adaptors with a motor protein that propels sequencing. Once the library is prepared, MinION sequencing can be carried out, allowing direct RNA sequencing from the 3′ poly(A) tail to the 5′ cap. The initial research has shown that dRNA-seq produces read lengths averaging around 1000 base pairs, with some reads exceeding 10,000 base pairs (111,113,124). Long reads have numerous benefits compared to short reads. They can enhance the detection of isoforms and estimate poly(A) tail length, which is crucial for alternative poly(A) analysis.

Regarding the advantages and disadvantages of long-read methods, nanopore direct-read RNA-seq is quite similar. However, direct-read technology has a unique advantage over other methods as it can detect epigenetic modifications to RNA, like methylation, and reduce bias. This technology is highly beneficial for researchers looking to analyze RNA sequences (111,125,126).

12.4.4 Single-Cell RNA-Seq

Conventional sequencing protocols on templates derived from a mixture of cells fail to detect the genomic variations present in each cell. Researchers rely on single cell sequencing of DNA or RNA to examine new biological inquiries. To perform single-cell sequencing, the first step is to isolate individual cells using various methods, including physical manipulation like micro pipetting in tubes, microfluidic techniques such as flow sorting or droplet capture, electric fields, and optical approaches such as optical tweezers. These techniques involve the distribution of cells into nanowells and in situ barcoding of cells. Once the library has been created, the NGS platform is used to sequence the template. The relatively new technology of single-cell RNA-Seq (scRNA-Seq) is gaining popularity, as it allows for capturing individual cells' distinct gene expression signatures, as opposed to the average abundances reported by bulk RNA-Seq methods. In transcriptomics, scRNA-Seq has proven helpful for examining gene interaction networks, lineage tracing, and rare cell populations, with numerous applications in the study of immunology and tumor heterogeneity (117,127).

12.5 RNA THERAPEUTICS AND SEQUENCING FOR COVID-19

The global outbreak of COVID-19 has resulted in considerable illness and fatalities, emphasizing the necessity for successful treatments. RNA therapeutics and sequencing have shown potential for creating effective COVID-19 treatments. For several years, extensive research has focused on utilizing mRNA as a therapeutic tool, particularly in developing vaccines for infectious diseases and some forms of cancer. One significant benefit of mRNA technology is the ability to quickly develop and manufacture treatments for newly emerging pathogens, as illustrated by the SARS-CoV-2 vaccines. Two vaccines developed with modified mRNA technology targeting SARS-CoV-2 have progressed rapidly through clinical trials. The mRNA vaccines developed by different manufacturers utilize lipid nanoparticles (LNPs) to ensure safety against Rnases. Both vaccines contain non-replicating mRNA sequences that code for the whole S-protein. However, they require different storage conditions due to different lipid multi-particulate (LMP) drug-delivery molecules. The Moderna COVID-19 vaccine must be stored at −20°C, whereas the BioNTech/Pfizer vaccine requires a much colder storage temperature of −80°C. Despite this difference, the shelf life of both vaccines is similar, lasting up to 6 months. The variation in storage conditions is mainly due to additional safety measures taken by BioNTech/Pfizer. The formulations of both vaccines contain identical active ingredients, vectors, and pH levels, with some differences in the excipients used (128,129). ASO technology can target mRNA, small RNA, or long non-coding RNA with no information being omitted. ASO therapy is highly advantageous as it can be applied against any viral RNA that is of interest. This emerging method has proven successful in treating various diseases such as myopathies, neurodegenerative diseases, oncology, and many others (26,130). ASO technologies have been developed for SARS-CoV to lower the severity of infection, identify the virus in human samples, and treat and diagnose associated diseases. Three patents have been published that use ASO for studying the pathogenesis and virulence of SARS-CoV. In one of these patents, WO2005023083, Ionis Pharmaceuticals describes an ASO that targets various regions of viral RNA to reduce the activity of the SARS virus and prevent or treat SARS virus-associated disease. The study uses a hybrid DNA/RNA ASO to disrupt the pseudoknot in the frameshift site of SARS-CoV. Ribozymes are RNA molecules with catalytic abilities by binding and breaking phosphodiester bonds in a targeted nucleic acid. In 2007, Japanese scientists designed a DNA/RNA chimeric ribozyme that can be used to treat general coronavirus; it is protected by a patent (JP2007043942). This chimeric ribozyme targets and cuts the conserved regions in the coronavirus family and can be adjusted for SARS-CoV-2 treatment strategies and research. Combining aptamers with ribozymes allows aptazymes—catalytic molecules—to be created. Researchers have found that when aptazymes are used with the CRISPR/Cas system as gRNA, they can significantly improve gene editing efficiency (131). Aptamer technology can potentially aid in both the diagnosis and treatment of COVID-19 by binding to antigenic viral proteins. Pinpoint Science has developed a nanosensor utilizing aptamers that has the potential to detect COVID-19 quickly. In addition, a comprehensive project proposal has been created outlining the application of aptamers in treating COVID-19 (132,133). SiRNAs can be utilized to silence the genes that encode structural and non-structural proteins of SARS-CoV-2. The viral genome of SARS-CoV-2 contains four vital structural proteins—envelope (E), membrane (M), nucleocapsid (N), and spike (S) proteins—which play a significant role in the assembly of the virion and obstructing viral replication in humans. The development of siRNAs has targeted these viral structural proteins' encoding genes for drug development (134). Feng Zhang's team introduced the SHERLOCK nucleic acid detection technique in 2017, the first of its kind that utilized CRISPR. This advanced method can identify RNA or DNA from clinically significant samples with high sensitivity, portability, and multiplexing capabilities. Metsky has created a tool for designing assays and a research approach that can detect SARS-CoV-2 and 66 related viruses. These details are outlined in a preprint on bioRxiv (135). The progress of advanced sequencing technologies, particularly NGS, has paved the way for RNA-seq to explore the extensive range of RNA species. This development is an exciting and apparent application that presents new prospects for enhancing the diagnosis and treatment of human ailments. RNA-seq is a comprehensive approach that offers a profound insight into the transcriptome and identifies innovative variants in RNA transcripts (136). Kim and colleagues conducted a thorough investigation utilizing the "sequencing-by-synthesis" and direct RNA sequencing approach to analyze a detailed map of the SARS-CoV-2 transcriptome. The results showed that this virus produces RNAs that encode unidentified ORFs and has at least 41 potential RNA modification sites. This study comprehensively explains the SARS-CoV-2 transcriptome, providing valuable insights into its genetic makeup (77). The term *transcriptome* refers to all the transcripts produced during a specific physiological state, such as when an individual is infected with SARS-CoV-2. In this case, the viral transcriptome is intertwined with the transcriptome of the host organism, as the transcriptional machinery enables the replication of SARS-CoV-2 viral particles (137,138). Over the past few months, experts have utilized single-cell RNA-seq to enhance our knowledge of COVID-19 concerning the ongoing worldwide crisis. By utilizing the scRNAseq technique, researchers have identified intricate and uncommon cell populations, uncovered regulatory connections between genes, and monitored the paths of specific cell lineages during development (127). Most RNA-seq data has been acquired indirectly due to the current techniques for analyzing transcriptomes, which involve converting RNA into cDNA before sequencing. However, this step can introduce various biases and artifacts that hinder transcripts' accurate characterization and quantification (139). Direct RNA sequencing effectively investigates the transcriptional layout of intricate genome viruses such as SARS-CoV-2. SARS-CoV-2 genome sequencing is being conducted in various settings, ranging from underdeveloped to highly developed countries. The analysis of these genomes has proven and will continue to prove invaluable in guiding public health measures related to COVID-19 and should therefore be sustained. In combating COVID-19, RNA therapeutics and sequencing have been indispensable assets. These technologies have enabled researchers to create successful vaccines and therapies, monitor the transmission of the virus, and detect novel strains—all of which are crucial in managing the pandemic.

12.6 CONCLUSION AND FUTURE DIRECTIONS

Recent reports on the development of RNA-based drugs have demonstrated the immense potential of this field. With the current knowledge, it is anticipated that new RNA-based drugs will emerge, offering treatments for diseases without existing methods. The number of RNA drugs currently in development and clinical trials is expanding rapidly. This growth can be attributed to resolving stability, delivery, and immunogenicity issues. The design and chemistry used to synthesize RNA aptamers, siRNAs, ASOs, and mRNAs have advanced to enable adequate stability and immune evasion while maintaining efficacy and specificity. Additionally, the discovery of potent and bio-compatible materials, aided by high-throughput screening technologies, has dramatically improved delivery technologies. RNA-seq can potentially transform clinical testing for many diseases due to its unique ability to detect global gene transcript levels and diverse RNA species simultaneously. This unprecedented capability of RNA-seq can usher in a new era of clinical testing.

The COVID-19 outbreak has highlighted the importance of innovative strategies in preventing and treating infectious diseases. The conventional methods of developing vaccines and discovering drugs have been challenging and time-consuming, resulting in many populations being at risk. Nevertheless, the emergence of RNA-based treatments and sequencing technologies has transformed infectious disease management.

REFERENCES

1. Gros François, Howard Hiatt, Walter Gilbert, Chuck G. Kurland, R. W. Risebrough, JDW. Unstable ribonucleic acid revealed by pulse labelling of *Escherichia coli*. *Nat.* 1961;190; 581–585.
2. Zamecnik PC, Stephenson ML. Inhibition of Rous sarcoma virus replication and cell transformation by a specific oligodeoxynucleotide. *Proceedings of the National Academy of Sciences of the United States of America* [Internet] 1978 [cited 2023 Apr 2];75(1); 280–284. Available from: www.pnas.org/doi/abs/10.1073/pnas.75.1.280
3. Wolff JA, Malone RW, Williams P, Chong W, Acsadi G, Jani A, et al. Direct gene transfer into mouse muscle in vivo. *Science* (80-) [Internet] 1990 [cited 2023 Apr 2];247(4949); 1465–1468. Available from: www.science.org/doi/10.1126/science.1690918
4. Martinon F, Krishnan S, Lenzen G, Magné R, Gomard E, Guillet JG, et al. Induction of virus-specific cytotoxic T lymphocytes in vivo by liposome-entrapped mRNA. *European Journal of Immunology* [Internet] 1993[cited 2023 Apr 2];23(7); 1719–1722. Available from: https://onlinelibrary.wiley.com/doi/full/10.1002/eji.1830230749
5. . . . BR-IA of P in A, 1998 Undefined. Fomivirsen approved for CMV retinitis. *pubmed.ncbi.nlm.nih.gov* [Internet] [cited 2023 Apr 2]; Available from: https://pubmed.ncbi.nlm.nih.gov/11365956/
6. Van Roessel P, Brand AH. Potent and specific genetic interference by double-stranded RNA in caenorhabditis elegans. *nature.com* [Internet] 2004 [cited 2023 Apr 2]. Available from: www.nature.com/articles/35888
7. McCaffrey AP, Meuse L, Pham TTT, Conklin DS, Hannon GJ, Kay MA. RNA interference in adult mice. *Nat.* 2002 [Internet] 2002 [cited 2023 Apr 2];418(6893); 38–39. Available from: www.nature.com/articles/418038a

8. Gragoudas ES, Adamis AP, Cunningham ET, Feinsod M, Guyer DR. Pegaptanib for neovascular age-related macular degeneration. *New England Journal of Medicine* [Internet] 2004 [cited 2023 Apr 2];351(27); 2805–2816. Available from: www.nejm.org/doi/abs/10.1056/NEJMoa042760

9. Weide B, Pascolo S, . . . BS-J of, 2009 undefined. Direct injection of protamine-protected mRNA: Results of a phase 1/2 vaccination trial in metastatic melanoma patients. *journals.lww.com* [Internet] [cited 2023 Apr 2]. Available from: https://journals.lww.com/immunotherapy-journal/Fulltext/2009/06000/Results_of_the_First_Phase_I_II_Clinical.8.aspx

10. Davis ME, Zuckerman JE, Choi CHJ, Seligson D, Tolcher A, Alabi CA, et al. Evidence of RNAi in humans from systemically administered siRNA via targeted nanoparticles. *Nat.* 2010 [Internet] 2010 [cited 2023 Apr 2];464(7291); 1067–1070. Available from: www.nature.com/articles/nature08956

11. Adams D, . . . AG-D-N england journal, 2018 undefined. Patisiran, an RNAi therapeutic, for hereditary transthyretin amyloidosis. *Massachusetts Medical Society* [Internet] [cited 2023 Apr 2]. Available from: www.nejm.org/doi/full/10.1056/nejmoa1716153

12. Kim YK. RNA therapy: Rich history, various applications and unlimited future prospects. *Experimental & Molecular Medicine* 2022 544 [Internet] 2022 [cited 2023 Apr 3];54(4); 455–465. Available from: www.nature.com/articles/s12276-022-00757-5

13. Bennett C. ES-A review of pharmacology, 2010 undefined. RNA targeting therapeutics: Molecular mechanisms of antisense oligonucleotides as a therapeutic platform. *annualreviews.org* [Internet] [cited 2023 Apr 2]. Available from: www.annualreviews.org/doi/abs/10.1146/annurev.pharmtox.010909.105654

14. Wu H, Lima WF, Zhang H, Fan A, Sun H, Crooke ST. Determination of the role of the human Rnase H1 in the pharmacology of DNA-like antisense drugs. *Journal of Biological Chemistry* [Internet] 2004 [cited 2023 Apr 2];279(17); 17181–17189. Available from: www.jbc.org/article/S0021925819755409/fulltext

15. Baker B, Lot S, . . . TC-J of B, 1997 undefined. 2′-O-(2-Methoxy) ethyl-modified anti-intercellular adhesion molecule 1 (ICAM-1) oligonucleotides selectively increase the ICAM-1 mRNA level and inhibit. *American Society for Biochemistry and Molecular Biology* [Internet] [cited 2023 Apr 2]. Available from: www.jbc.org/article/S0021-9258(18)40547-9/abstract

16. Larrouy B, Blonski C, Boiziau C, Stuer M, Gene SM, 1992 Undefined. Rnase H-mediated inhibition of translation by antisense oligodeoxyribo-nucleotides: Use of backbone modification to improve specificity. *Elsevier* [Internet] [cited 2023 Apr 2]. Available from: www.sciencedirect.com/science/article/pii/0378111992901215

17. McClorey G, pharmacology MW-C Opinion in, 2015 Undefined. An overview of the clinical application of antisense oligonucleotides for RNA-targeting therapies. *Elsevier* [Internet] [cited 2023 Apr 2]. Available from: www.sciencedirect.com/science/article/pii/S1471489215000922

18. Rossor AM, Reilly MM, Sleigh JN. Antisense oligonucleotides and other genetic therapies made simple. *Practical Neurology* [Internet] 2018 [cited 2023 Apr 2];18(2); 126–131. Available from: https://pn.bmj.com/content/18/2/126

19. Vickers TA, Wyatt JR, Burckin T, Bennett CF, Freier SM. Fully modified 2′ MOE oligonucleotides redirect polyadenylation. *Nucleic Acids Research* 2001;29(6); 1293–1299.

20. Liang XH, Shen W, Sun H, Migawa MT, Vickers TA, Crooke ST. Translation efficiency of mRNAs is increased by antisense oligonucleotides targeting upstream open reading frames. *Nature Biotechnology* [Internet] [cited 2023 Apr 2] 2016;34(8); 875–880. Available from: www.nature.com/articles/nbt.3589

21. Hua Y, Vickers TA, Baker BF, Bennett CF, Krainer AR. Enhancement of SMN2 exon 7 inclusion by antisense oligonucleotides targeting the exon. *PLoS Biology* 2007;5(4); 729–744.

22. Lim KRQ, Maruyama R, Yokota T. Eteplirsen in the treatment of Duchenne muscular dystrophy. *Drug Design, Development and Therapy* [Internet] 2017 [cited 2023 Apr 2];11; 533–545. Available from: http://dx.doi.org/10.2147/DDDT.S97635

23. Neil EE, Bisaccia EK. Nusinersen: A novel antisense oligonucleotide for the treatment of spinal muscular atrophy. *Journal of Pediatric Pharmacology and Therapeutics* [Internet] 2019 [cited 2023 Apr 2];24(3); 194–203. Available from: https://meridian.allenpress.com/jppt/article/24/3/194/433570/Nusinersen-A-Novel-Antisense-Oligonucleotide-for

24. Mathew V, design AW-D, Therapy Development and, 2019 Undefined. Inotersen: New promise for the treatment of hereditary transthyretin amyloidosis. *Taylor & Francis* [Internet] 2022 [cited 2023 Apr 2];13; 1515–1525. Available from: www.tandfonline.com/doi/abs/10.2147/DDDT.S162913

25. Godfrey C, Desviat LR, Smedsrød B, Piétri-Rouxel F, Denti MA, Disterer P, et al. Delivery is key: Lessons learnt from developing splice-switching antisense therapies. *EMBO Molecular Medicine*

[Internet] 2017 [cited 2023 Apr 2];9(5); 545–557. Available from: https://onlinelibrary.wiley.com/doi/full/10.15252/emmm.201607199

26. Sardone V, Zhou H, Muntoni F, Ferlini A, Falzarano MS, Shiu W, et al. Antisense oligonucleotide-based therapy for neuromuscular disease. *mdpi.com* [Internet] [cited 2023 Apr 2]. Available from: www.mdpi.com/190002

27. Quemener AM, Galibert MD. Antisense oligonucleotide: A promising therapeutic option to beat COVID-19. *Wiley Interdisciplinary Reviews RNA* [Internet] 2022 [cited 2023 May 4];13(4); e1703. Available from: https://onlinelibrary.wiley.com/doi/full/10.1002/wrna.1703

28. Chery J. RNA therapeutics: RNAi and antisense mechanisms and clinicalapplications. *Postdoc J a J Postdr Res Postdr Aff.* [Internet] 2016 [cited 2023 Apr 2];4(7); 35. Available from: /pmc/articles/PMC4995773/

29. Geary R, Norris D, Yu R, Reviews CB-A Drug Delivery, 2015 Undefined. Pharmacokinetics, biodistribution and cell uptake of antisense oligonucleotides. *Elsevier* [Internet] [cited 2023 Apr 2]. Available from: www.sciencedirect.com/science/article/pii/S0169409X15000101

30. Ozcan G, Ozpolat B, Coleman R, . . . AS-A Drug Delivery, 2015 Undefined. Preclinical and clinical development of siRNA-based therapeutics. *Elsevier* [Internet] [cited 2023 Apr 2]. Available from: www.sciencedirect.com/science/article/pii/S0169409X15000095

31. Jorgensen R. Altered gene expression in plants due to trans interactions between homologous genes. *Trends in Biotechnology* [Internet] 1990 [cited 2023 Apr 2];8(12); 340–344. Available from: https://europepmc.org/article/med/1366894

32. Sen GL, Blau HM. A brief history of RNAi: The silence of the genes. *The FASEB Journal* 2006;20(9); 1293–1299.

33. Wittrup A, Lieberman J. Knocking down disease: a progress report on siRNA therapeutics. *Nature Reviews Genetics* 2015 [Internet] [cited 2023 Apr 2];16(9); 543–552. Available from: www.nature.com/articles/nrg3978

34. Molecular mechanisms and biological functions of siRNA. *ncbi.nlm.nih.gov* [Internet]. [cited 2023 Apr 2]. Available from: www.ncbi.nlm.nih.gov/pmc/articles/PMC5542916/

35. Sardh E, Harper P, Balwani M, Stein P, Rees D, Bissell DM, et al. Phase 1 trial of an rna interference therapy for acute intermittent porphyria. *New England Journal of Medicine* 2019;380(6); 549–558.

36. Snøve O, Holen T. Many commonly used siRNAs risk off-target activity. *Biochemical and Biophysical Research Communications* 2004;319(1); 256–263.

37. Doench JG, Petersen CP, Sharp PA. siRNAs can function as miRNAs. *Genes & Development* [Internet] 2003 [cited 2023 Apr 2];17(4); 438–442. Available from: http://genesdev.cshlp.org/content/17/4/438.full

38. Biology JL-N Structural & Molecular, 2018 Undefined. Tapping the RNA world for therapeutics. *nature.com* [Internet] [cited 2023 Apr 2]. Available from: www.nature.com/articles/s41594-018-0054-4

39. Gilleron J, Querbes W, Zeigerer A, Borodovsky A, Marsico G, Schubert U, et al. Image-based analysis of lipid nanoparticle–mediated siRNA delivery, intracellular trafficking and endosomal escape. *Nature Biotechnology* 2013 [Internet] [cited 2023 Apr 2];31(7); 638–646. Available from: www.nature.com/articles/nbt.2612

40. Wittrup A, Ai A, Liu X, Hamar P, . . . RT-N, 2015 undefined. Visualizing lipid-formulated siRNA release from endosomes and target gene knockdown. *nature.com* [Internet] [cited 2023 Apr 2]. Available from: www.nature.com/articles/nbt.3298

41. Bartlett DW, Davis ME. Insights into the kinetics of siRNA-mediated gene silencing from live-cell and live-animal bioluminescent imaging. *Nucleic Acids Research* [Internet] 2006 [cited 2023 Apr 2];34(1); 322–333. Available from: https://academic.oup.com/nar/article/34/1/322/2401639

42. Jinek M, Nature JD, 2009 Undefined. A three-dimensional view of the molecular machinery of RNA interference. *nature.com* [Internet] [cited 2023 Apr 2]. Available from: www.nature.com/articles/nature07755

43. Lewis BP, Burge CB, Bartel DP. Conserved seed pairing, often flanked by adenosines, indicates that thousands of human genes are microRNA targets. *Cell* 2005;120(1); 15–20.

44. Lee RC, Feinbaum RL, Ambros V. The *C. elegans* heterochronic gene lin-4 encodes small RNAs with antisense complementarity to lin-14. *Cell* 1993;75(5); 843–854.

45. Lam JKW, Chow MYT, Zhang Y, Leung SWS. siRNA versus miRNA as therapeutics for gene silencing. *Molecular Therapy Nucleic Acids* [Internet] 2015 [cited 2023 Apr 2];4(9); e252. Available from: https://pubmed.ncbi.nlm.nih.gov/26372022/

46. Lou S, Sun T, Li H, Hu Z. Mechanisms of microRNA-mediated gene regulation in unicellular model alga *Chlamydomonas reinhardtii*. *Biotechnology for Biofuels* 2018;11(1).

47. O'Brien J, Hayder H, Zayed Y, Peng C. Overview of microRNA biogenesis, mechanisms of actions, and circulation. *Frontiers in Endocrinology* (Lausanne) 2018;9(AUG).

48. Elmén J, Lindow M, Schütz S, Lawrence M, Nature AP-, 2008 undefined. LNA-mediated microRNA silencing in non-human primates. *nature.com* [Internet] [cited 2023 Apr 2]. Available from: www.nature.com/articles/nature06783

49. Obad S, Dos Santos CO, Petri A, Heidenblad M, Broom O, Ruse C, et al. Silencing of microRNA families by seed-targeting tiny LNAs. *Nature Genetics* 2011 [Internet] [cited 2023 Apr 2];43(4); 371–378. Available from: www.nature.com/articles/ng.786

50. Janssen HLA, Reesink HW, Lawitz EJ, Zeuzem S, Rodriguez-Torres M, Patel K, et al. Treatment of HCV infection by targeting microRNA. *New England Journal of Medicine* 2013;368(18); 1685–1694.

51. Ellington A, Nature JS, 1990 Undefined. In vitro selection of RNA molecules that bind specific ligands. *nature.com* [Internet] [cited 2023 Apr 2]. Available from: www.nature.com/articles/346818a0

52. Tuerk C, Gold L. Systematic evolution of ligands by exponential enrichment: RNA ligands to bacteriophage T4 DNA polymerase. *Science* (80-) [Internet] 1990 [cited 2023 Apr 2];249(4968); 505–510. Available from: www.science.org/doi/10.1126/science.2200121

53. Chemistry SJ-C, 1999 undefined. Aptamers: An emerging class of molecules that rival antibodies in diagnostics. *academic.oup.com* [Internet] 1999 [cited 2023 Apr 2]. Available from https://academic.oup.com/clinchem/article-abstract/45/9/1628/5643446

54. Yu Y, Liang C, Lv Q, Li D, Xu X, Liu B, et al. Molecular selection, modification and development of therapeutic oligonucleotide aptamers. *International Journal of Molecular Sciences* 2016 [Internet] 2016 [cited 2023 Apr 2];17(3); 358. Available from: www.mdpi.com/1422-0067/17/3/358/htm

55. Cruz-Toledo J, McKeague M, Zhang X, Giamberardino A, McConnell E, Francis T, et al. Aptamer base: A collaborative knowledge base to describe aptamers and SELEX experiments. *Database* [Internet]. 2012 [cited 2023 Apr 2]; 2012. Available from: https://academic.oup.com/database/article/doi/10.1093/database/bas006/431883

56. Nature WG, 1986 Undefined. Origin of life: The RNA world. *nature.com* [Internet] [cited 2023 Apr 2]. Available from: www.nature.com/articles/319618a0

57. Wright M, Conry RM, Lobuglio AF, Wright M, Sumerel L, Pike MJ, et al. Characterization of a messenger RNA polynucleotide vaccine vector. *AACR* [Internet] 1397 [cited 2023 Apr 2]; 55. Available from: https://aacrjournals.org/cancerres/article-abstract/55/7/1397/502087

58. Sahin U, Karikó K, discovery ÖT-N reviews D, 2014 Undefined. mRNA-based therapeutics—Developing a new class of drugs. *nature.com* [Internet] [cited 2023 Apr 2]. Available from: www.nature.com/articles/nrd4278

59. Pardi N, Hogan MJ, Porter FW, Weissman D. mRNA vaccines—A new era in vaccinology. *Nature Reviews Drug Discovery* 2018 [Internet] [cited 2023 Apr 2];17(4); 261–279. Available from: www.nature.com/articles/nrd.2017.243

60. Heiser A, Coleman D, Dannull J, Yancey D, Maurice MA, Lallas CD, et al. Autologous dendritic cells transfected with prostate-specific antigen RNA stimulate CTL responses against metastatic prostate tumors. *Journal of Clinical Investigation* 2002;109(3); 409–417.

61. Morse MA, Nair SK, Boczkowski D, Tyler D, Hurwitz HI, Proia A, et al. The feasibility and safety of immunotherapy with dendritic cells loaded with CEA mRNA following neoadjuvant chemoradiotherapy and resection of pancreatic cancer. *Journal of Gastrointestinal Cancer* 2002;32(1); 1–6.

62. Morse MA, Nair SK, Mosca PJ, Hobeika AC, Clay TM, Deng Y, et al. Immunotherapy with autologous, human dendritic cells transfected with carcinoembryonic antigen mRNA. *Cancer Investigation* 2003;21(3); 341–349.

63. Rittig SM, Haentschel M, Weimer KJ, Heine A, Muller MR, Brugger W, et al. Intradermal vaccinations with RNA coding for TAA generate CD8+ and CD4+ immune responses and induce clinical benefit in vaccinated patients. *Molecular Therapy* 2011;19(5); 990–999.

64. Kübler H, Maurer T, Stenzl A, Feyerabend S, Steiner U, Schostak M, et al. Final analysis of a phase I/IIa study with CV9103, an intradermally administered prostate cancer immunotherapy based on self-adjuvanted mRNA. *Journal of Clinical Oncology* 2011;29(Suppl 15); 4535–4535. Available from: https://doi.org/101200/jco20112915_suppl4535.

65. Sebastian M, von Boehmer L, Zippelius A, Mayer F, Reck M, Atanackovic D, et al. Messenger RNA vaccination and B-cell responses in NSCLC patients. *Journal of Clinical Oncology* 2012;30(Suppl 15); 2573–2573.

66. Wilgenhof S, Van Nuffel AMT, Benteyn D, Corthals J, Aerts C, Heirman C, et al. A phase IB study on intravenous synthetic mRNA electroporated dendritic cell immunotherapy in pretreated advanced melanoma patients. *Annals of Oncology* 2013;24(10); 2686–2693.

67. Karikó K, Muramatsu H, Welsh F, Ludwig J, Therapy HK-M, 2008 Undefined. Incorporation of pseudouridine into mRNA yields superior nonimmunogenic vector with increased translational capacity and biological stability. *Elsevier* [Internet] [cited 2023 Apr 2]. Available from: www.sciencedirect.com/science/article/pii/S1525001616326818

68. Kauffman KJ, Webber MJ, Anderson DG. Materials for non-viral intracellular delivery of messenger RNA therapeutics. *Journal of Controlled Release* 2016;240; 227–234.

69. Guan S, Rosenecker J. Nanotechnologies in delivery of mRNA therapeutics using nonviral vector-based delivery systems. *Gene Therapy* 2017 [Internet] 2017[cited 2023 Apr 2];24(3); 133–143. Available from: www.nature.com/articles/gt20175

70. Thess A, Grund S, Mui BL, Hope MJ, Baumhof P, Fotin-Mleczek M, et al. Sequence-engineered mRNA without chemical nucleoside modifications enables an effective protein therapy in large animals. *Molecular Therapy* 2015;23(9); 1456–1464.

71. Karikó K, Muramatsu H, . . . JL-N Acids, 2011 Undefined. Generating the optimal mRNA for therapy: HPLC purification eliminates immune activation and improves translation of nucleoside-modified, protein-encoding mRNA. *academic.oup.com* [Internet] [cited 2023 Apr 2]; Available from: https://academic.oup.com/nar/article-abstract/39/21/e142/1104771

72. Unniyampurath U, Pilankatta R, Krishnan MN. RNA Interference in the age of CRISPR: Will CRISPR Interfere with RNAi? *Int J Mol Sci* 2016, Vol 17, Page 291 [Internet]. 2016 Feb 26 [cited 2023 Apr 2];17(3); 291. Available from: www.mdpi.com/1422-0067/17/3/291/htm

73. Barrangou R, Fremaux C, Deveau H, Richards M, Boyaval P, Moineau S, et al. CRISPR provides acquired resistance against viruses in prokaryotes. *Science* (80-) [Internet] 2007 [cited 2023 Apr 2];315(5819); 1709–1712. Available from: www.science.org/doi/10.1126/science.1138140

74. Sorek R, Kunin V, Microbiology PH-NR, 2008 undefined. CRISPR—a widespread system that provides acquired resistance against phages in bacteria and archaea. *nature.com* [Internet] [cited 2023 Apr 2]. Available from: www.nature.com/articles/nrmicro1793

75. Grissa I, Vergnaud G, Pourcel C. The CRISPRdb database and tools to display CRISPRs and to generate dictionaries of spacers and repeats. *BMC Bioinformatics* [Internet] 2007 [cited 2023 Apr 2];8(1); 1–10. Available from: https://link.springer.com/articles/10.1186/1471-2105-8-172

76. Mali P, Yang L, Esvelt KM, Aach J, Guell M, DiCarlo JE, et al. RNA-guided human genome engineering via Cas9. *Science* (80-) 2013;339(6121); 823–826.

77. Makarova KS, Wolf YI, Alkhnbashi OS, Costa F, Shah SA, Saunders SJ, et al. An updated evolutionary classification of CRISPR–Cas systems. *Nature Reviews Microbiology* 2015 [Internet] [cited 2023 Apr 2];13(11); 722–736. Available from: www.nature.com/articles/nrmicro3569

78. Shalem O, Sanjana NE, Zhang F. High-throughput functional genomics using CRISPR–Cas9. *Nature Reviews Genetics* 2015 [Internet]. 2015 [cited 2023 Apr 2];16(5); 299–311. Available from: www.nature.com/articles/nrg3899

79. Choudhary E, Thakur P, Pareek M, Communications NA-N, 2015 Undefined. Gene silencing by CRISPR interference in mycobacteria. *nature.com* [Internet] [cited 2023 Apr 2]. Available from: www.nature.com/articles/ncomms7267

80. Singh A, Carette X, Potluri L, . . . JS-N Acids, 2016 Undefined. Investigating essential gene function in *Mycobacterium tuberculosis* using an efficient CRISPR interference system. *academic.oup.com* [Internet] [cited 2023 Apr 2]. Available from: https://academic.oup.com/nar/article-abstract/44/18/e143/2468349

81. Heald AE, Iversen PL, Saoud JB, Sazani P, Charleston JS, Axtelle T, et al. Safety and pharmacokinetic profiles of phosphorodiamidate morpholino oligomers with activity against Ebola virus and Marburg virus: Results of two single-ascending-dose studies. *Antimicrobial Agents and Chemotherapy* [Internet] 2014 [cited 2023 May 4];58(11); 6639–6647. Available from: https://journals.asm.org/doi/10.1128/AAC.03442-14

82. DeJesus E, Harward S, Jewell RC, Johnson M, Dumont E, Wilches V, et al. A phase IIa study evaluating safety, pharmacokinetics, and antiviral activity of GSK2838232, a novel, second-generation maturation inhibitor, in participants with human immunodeficiency virus type 1 infection. *Clinical Infectious Diseases* [Internet] 2020 [cited 2023 May 4];71(5); 1255–1262. Available from: https://clinicaltrials.gov/ct2/show/NCT03045861

83. Beigel JH, Voell J, Muñoz P, Kumar P, Brooks KM, Zhang J, et al. Safety, tolerability, and pharmacokinetics of radavirsen (AVI-7100), an antisense oligonucleotide targeting influenza a M1/M2 translation.

British Journal of Clinical Pharmacology [Internet] 2018 [cited 2023 May 4];84(1); 25–34. Available from: https://onlinelibrary.wiley.com/doi/full/10.1111/bcp.13405

84. Dunning J, Sahr F, Rojek A, Gannon F, Carson G, Idriss B, et al. Experimental treatment of Ebola virus disease with TKM-130803: A single-arm phase 2 clinical trial. *PLoS Medicine* [Internet] 2016 [cited 2023 May 4];13(4). Available from: https://pubmed.ncbi.nlm.nih.gov/27093560/

85. Khaitov M, Nikonova A, Kofiadi I, Shilovskiy I, Smirnov V, Elisytina O, et al. Treatment of COVID-19 patients with a SARS-CoV-2-specific siRNA-peptide dendrimer formulation. *Allergy* [Internet] 2023 [cited 2023 May 4]. Available from: https://onlinelibrary.wiley.com/doi/full/10.1111/all.15663

86. van der Ree MH, de Vree JM, Stelma F, Willemse S, van der Valk M, Rietdijk S, et al. Safety, tolerability, and antiviral effect of RG-101 in patients with chronic hepatitis C: A phase 1B, double-blind, randomised controlled trial. *Lancet* 2017;389(10070); 709–717.

87. Jang KJ, Lee NR, Yeo WS, Jeong YJ, Kim DE. Isolation of inhibitory RNA aptamers against severe acute respiratory syndrome (SARS) coronavirus NTPase/Helicase. *Biochemical and Biophysical Research Communications* 2008;366(3); 738–744.

88. Shubham S, Hoinka J, Banerjee S, Swanson E, Dillard JA, Lennemann NJ, et al. A 2'FY-RNA Motif Defines an Aptamer for Ebolavirus Secreted Protein [cited 2023 May 4]. Available from: www.nature.com/scientificreports/

89. Kensch O, Connolly BA, Steinhoff HJ, McGregor A, Goody RS, Restle T. Hiv-1 reverse transcriptase-pseudoknot RNA aptamer interaction has a binding affinity in the low picomolar range coupled with high specificity. *Journal of Biological Chemistry* [Internet] 2000 [cited 2023 May 4];275(24); 18271–18278. Available from: www.jbc.org/article/S0021925819831824/fulltext

90. Chaloin L, Lehmann MJ, Sczakiel G, Restle T. Endogenous expression of a high-affinity pseudoknot RNA aptamer suppresses replication of HIV-1. *Nucleic Acids Research* [Internet] 2002 [cited 2023 May 4];30(18); 4001–4008. Available from: https://academic.oup.com/nar/article/30/18/4001/1075449

91. Li N, Wang Y, Pothukuchy A, Syrett A, Husain N, Gopalakrisha S, et al. Aptamers that recognize drug-resistant HIV-1 reverse transcriptase. *Nucleic Acids Research* [Internet] 2008 [cited 2023 May 4];36(21); 6739–6751. Available from: https://academic.oup.com/nar/article/36/21/6739/2410436

92. Whatley AS, Ditzler MA, Lange MJ, Biondi E, Sawyer AW, Chang JL, et al. Potent inhibition of HIV-1 reverse transcriptase and replication by nonpseudoknot, "UCAA-motif" RNA aptamers. *Molecular Therapy—Nucleic Acids* 2013;2; e71.

93. Ramalingam D, Duclair S, Datta SAK, Ellington A, Rein A, Prasad VR. RNA aptamers directed to human immunodeficiency virus type 1 gag polyprotein bind to the matrix and nucleocapsid domains and inhibit virus production. *Journal of Virology* [Internet] 2011 [cited 2023 May 4];85(1); 305–314. Available from: https://journals.asm.org/doi/10.1128/JVI.02626-09

94. Lee CH, Lee YJ, Kim JH, Lim JH, Kim JH, Han W, et al. Inhibition of hepatitis C virus (HCV) replication by specific RNA aptamers against HCV NS5B RNA replicase. *Journal of Virology* [Internet] 2013 [cited 2023 May 4];87(12); 7064–7074. Available from: https://journals.asm.org/doi/10.1128/JVI.00405-13

95. Marton S, Romero-López C, Berzal-Herranz A. RNA aptamer-mediated interference of HCV replication by targeting the CRE-5BSL3.2 domain. *Journal of Viral Hepatitis* [Internet]. 2013 [cited 2023 May 4];20(2); 103–112. Available from: https://onlinelibrary.wiley.com/doi/full/10.1111/j.1365-2893.2012.01629.x

96. Polack FP, Thomas SJ, Kitchin N, Absalon J, Gurtman A, Lockhart S, et al. Safety and efficacy of the BNT162b2 mRNA COVID-19 vaccine. *New England Journal of Medicine* [Internet] 2020 [cited 2023 May 4];383(27); 2603–2615. Available from: www.nejm.org/doi/full/10.1056/nejmoa2034577

97. Walsh EE, Frenck RW, Falsey AR, Kitchin N, Absalon J, Gurtman A, et al. Safety and immunogenicity of two RNA-Based COVID-19 vaccine candidates. *New England Journal of Medicine* [Internet] 2020 [cited 2023 May 4];383(25); 2439–2450. Available from: www.nejm.org/doi/full/10.1056/nejmoa2027906

98. Baden LR, El Sahly HM, Essink B, Kotloff K, Frey S, Novak R, et al. Efficacy and safety of the mRNA-1273 SARS-CoV-2 vaccine. *New England Journal of Medicine* [Internet] 2021 [cited 2023 May 4];384(5); 403–416. Available from: www.nejm.org/doi/full/10.1056/nejmoa2035389

99. Kremsner PG, Ahuad Guerrero RA, Arana-Arri E, Aroca Martinez GJ, Bonten M, Chandler R, et al. Efficacy and safety of the CvnCoV SARS-CoV-2 mRNA vaccine candidate in ten countries in Europe and Latin America (HERALD): A randomised, observer-blinded, placebo-controlled, phase 2b/3 trial. *Lancet Infectious Diseases* 2022;22(3);329–340.

100. Dorottya Laczkó A, Hogan MJ, Toulmin SA, Allman D, Locci M, Pardi Correspondence N, et al. A single immunization with nucleoside-modified mRNA vaccines elicits strong cellular and humoral immune responses against SARS-CoV-2 in Mice ll a single immunization with nucleoside-modified mRNA vaccines elicits strong cellular and humoral immune responses against SARS-CoV-2 in mice. *Immunity* [Internet] 2020 [cited 2023 May 4];53; 724–732; e7. Available from: https://doi.org/10.1016/j.immuni.2020.07.019

101. de Alwis R, Gan ES, Chen S, Leong YS, Tan HC, Zhang SL, et al. A single dose of self-transcribing and replicating RNA-based SARS-CoV-2 vaccine produces protective adaptive immunity in mice. *Molecular Therapy* 2021;29(6); 1970–1983.

102. Kalnin KV., Plitnik T, Kishko M, Zhang J, Zhang D, Beauvais A, et al. Immunogenicity and efficacy of mRNA COVID-19 vaccine MRT5500 in preclinical animal models. *npj Vaccines* 2021 [Internet] [cited 2023 May 4];6(1); 1–12. Available from: www.nature.com/articles/s41541-021-00324-5

103. Alberer M, Gnad-Vogt U, Hong HS, Mehr KT, Backert L, Finak G, et al. Safety and immunogenicity of a mRNA rabies vaccine in healthy adults: An open-label, non-randomised, prospective, first-in-human phase 1 clinical trial. *Lancet* 2017;390(10101); 1511–1520.

104. Aldrich C, Leroux–Roels I, Huang KB, Bica MA, Loeliger E, Schoenborn-Kellenberger O, et al. Proof-of-concept of a low-dose unmodified mRNA-based rabies vaccine formulated with lipid nanoparticles in human volunteers: A phase 1 trial. *Vaccine* 2021;39(8); 1310–1318.

105. Gay CL, Debenedette MA, Tcherepanova IY, Gamble A, Lewis WE, Cope AB, et al. Immunogenicity of AGS-004 dendritic cell therapy in patients treated during acute HIV infection. *https://home.liebertpub.com/aid* [Internet] 2018 [cited 2023 May 4];34(1); 111–122. Available from: www.liebertpub.com/doi/10.1089/aid.2017.0071

106. De Jong W, Aerts J, Allard S, Brander C, Buyze J, Florence E, et al. IHIVARNA phase IIa, a randomized, placebo-controlled, double-blinded trial to evaluate the safety and immunogenicity of iHIVARNA-01 in chronically HIV-infected patients under stable combined antiretroviral therapy. *Trials* 2019;20(1).

107. *Safety, Reactogenicity, and Immunogenicity of mRNA-1653 in Healthy Adults—Full Text View—ClinicalTrials.gov* [Internet]. [cited 2023 May 4]. Available from: https://clinicaltrials.gov/ct2/show/NCT03392389

108. Yuan M, Zhang W, Wang J, Yaghchi C Al, Ahmed J, Chard L, et al. Efficiently editing the vaccinia virus genome by using the CRISPR-Cas9 system. *Journal of Virology* [Internet] 2015 [cited 2023 Apr 2]. Available from: https://journals.asm.org/doi/10.1128/JVI.00339-15

109. Tang YD, Liu JT, Fang QQ, Wang TY, Sun MX, An TQ, et al. Recombinant pseudorabies virus (PRV) expressing firefly luciferase effectively screened for CRISPR/Cas9 single guide RNAs and anti-viral compounds. *Viruses* 2016 [Internet] [cited 2023 Apr 2]; 8(4); 90. Available from www.mdpi.com/1999-4915/8/4/90/htm

110. Methods SS-N, 2008 Undefined. Next-generation sequencing transforms today's biology. *nature.com* [Internet] 2008 [cited 2023 Apr 2];5(1). Available from: www.nature.com/articles/nmeth1156

111. Garalde D, Snell E, Jachimowicz D, . . . BS-N, 2018 Undefined. Highly parallel direct RNA sequencing on an array of nanopores. *nature.com* [Internet] [cited 2023 Apr 2]. Available from: www.nature.com/articles/nmeth.4577

112. Smith AM, Jain M, Mulroney L, Garalde DR, Akeson M. Reading canonical and modified nucleobases in 16S ribosomal RNA using nanopore native RNA sequencing. *PloS One* [Internet] 2019 [cited 2023 Apr 2];14(5); e0216709. Available from: https://journals.plos.org/plosone/article?id=10.1371/journal.pone.0216709

113. Byrne A, Beaudin AE, Olsen HE, Jain M, Cole C, Palmer T, et al. Nanopore long-read RNAseq reveals widespread transcriptional variation among the surface receptors of individual B cells. *Nature Communications* 2017 [Internet] [cited 2023 Apr 2];8(1); 1–11. Available from: www.nature.com/articles/ncomms16027

114. Hong M, Tao S, Zhang L, Diao LT, Huang X, Huang S, et al. RNA sequencing: New technologies and applications in cancer research. *Journal of Hematology & Oncology* 2020 [Internet] [cited 2023 May 5];13(1); 1–16. Available from: https://jhoonline.biomedcentral.com/articles/10.1186/s13045-020-01005-x

115. biotechnology N, 2014 undefined. A comprehensive assessment of RNA-seq accuracy, reproducibility and information content by the sequencing quality control consortium. *nature.com* [Internet] [cited 2023 Apr 2]. Available from: www.nature.com/articles/nbt.2957

116. Li S, Tighe SW, Nicolet CM, Grove D, Levy S, Farmerie W, et al. Multi-platform assessment of transcriptome profiling using RNA-seq in the ABRF next-generation sequencing study. *Nature*

Biotechnology [Internet] 2014 [cited 2023 Apr 3];32(9); 915–925. Available from: https://pubmed.ncbi.nlm.nih.gov/25150835/

117. Stark R, Grzelak M, Genetics JH-NR, 2019 Undefined. RNA sequencing: The teenage years. *nature.com* [Internet] [cited 2023 Apr 3]. Available from: www.nature.com/articles/s41576-019-0150-2

118. Sharon D, Tilgner H, Grubert F, Snyder M. A single-molecule long-read survey of the human transcriptome. *Nature Biotechnology* 2013 [Internet] [cited 2023 Apr 3];31(11); 1009–1014. Available from: www.nature.com/articles/nbt.2705

119. Cartolano M, Huettel B, Hartwig B, Reinhardt R, Schneeberger K. cDNA library enrichment of full length transcripts for SMRT long read sequencing. *PloS One* 2016;11(6).

120. Oikonomopoulos S, Wang YC, Djambazian H, Badescu D, Ragoussis J. Benchmarking of the Oxford Nanopore MinION sequencing for quantitative and qualitative assessment of cDNA populations. *Springer* [Internet] 2016 [cited 2023 Apr 3]. Available from: https://link.springer.com/content/pdf/10.1038/srep31602.pdf

121. Engström PG, Steijger T, Sipos B, Grant GR, Kahles A, Rätsch G, et al. Systematic evaluation of spliced alignment programs for RNA-seq data. *Nature Methods* [Internet] [cited 2023 Apr 3];10(12); 1185–1191. Available from: https://pubmed.ncbi.nlm.nih.gov/24185836/

122. Ardui S, Ameur A, . . . JV-N Acids, 2018 Undefined. Single molecule real-time (SMRT) sequencing comes of age: Applications and utilities for medical diagnostics. *academic.oup.com* [Internet] [cited 2023 Apr 3]. Available from: https://academic.oup.com/nar/article-abstract/46/5/2159/4833218

123. Jain M, Olsen HE, Paten B, Akeson M. The Oxford nanopore MinION: Delivery of nanopore sequencing to the genomics community. *Genome Biology* 2016;17(1).

124. Workman RE, Tang AD, Tang PS, Jain M, Tyson JR, Razaghi R, et al. Nanopore native RNA sequencing of a human poly(A) transcriptome. *Nature Methods* 2019 [Internet] [cited 2023 Apr 3];16(12); 1297–305. Available from: www.nature.com/articles/s41592-019-0617-2

125. Tilgner H, Grubert F, Sharon D, Snyder MP. Defining a personal, allele-specific, and single-molecule long-read transcriptome. *Proceedings of the National Academy of Sciences of the United States of America* 2014;111(27); 9869–9874.

126. Schwartz S, Motorin Y. Next-generation sequencing technologies for detection of modified nucleotides in RNAs. *RNA Biology* 2017;14(9); 1124–1137.

127. Hwang B, Lee JH, Bang D. Single-cell RNA sequencing technologies and bioinformatics pipelines. *Experimental & Molecular Medicine* 2018 [Internet] [cited 2023 Apr 3];50(8); 1–14. Available from: www.nature.com/articles/s12276-018-0071-8

128. Uddin M, Vaccines MR-, 2021 Undefined. Challenges of storage and stability of mRNA-based COVID-19 vaccines. *mdpi.com* [Internet] 2021 [cited 2023 Apr 3]. Available from: www.mdpi.com/2076-393X/9/9/1033

129. Tavilani A, Abbasi E, Ara F, Darini A, open ZA-M, 2021 Undefined. COVID-19 vaccines: Current evidence and considerations. *Elsevier* [Internet] [cited 2023 Apr 3]. Available from: www.sciencedirect.com/science/article/pii/S2589936821000487

130. Moreno PMD, Pêgo AP. Therapeutic antisense oligonucleotides against cancer: Hurdling to the clinic. *Frontiers in Chemistry* 2014;2(OCT).

131. Tang W, Hu JH, Liu DR. Aptazyme-embedded guide RNAs enable ligand-responsive genome editing and transcriptional activation. *Nature Communications* 2017 [Internet] [cited 2023 Apr 3];8(1); 1–8. Available from: www.nature.com/articles/ncomms15939

132. Song M-S, Lee S-W, Jung H, Ahn D-G, Jeon I-J, Kim JD, et al. RNA aptamer-based sensitive detection of SARS coronavirus nucleocapsid protein. *pubs.rsc.org* [Internet] 2009 [cited 2023 Apr 3]. Available from: https://pubs.rsc.org/en/content/articlehtml/2009/an/b906788d

133. Jang K, Lee N, Yeo W, Jeong Y, . . . DK Biophysical Research, 2008 Undefined. Isolation of inhibitory RNA aptamers against severe acute respiratory syndrome (SARS) coronavirus NTPase/Helicase. *Elsevier* [Internet] [cited 2023 Apr 3]. Available from: www.sciencedirect.com/science/article/pii/S0006291X07026319

134. Shi Y, Yang DH, Xiong J, Jia J, Huang B, Jin YX. Inhibition of genes expression of SARS coronavirus by synthetic small interfering RNAs. *Cell Research* 2005 [Internet] [cited 2023 Apr 3];15(3); 193–200. Available from: www.nature.com/articles/7290286

135. Metsky HC, Freije CA, Kosoko-Thoroddsen T-SF, Sabeti PC, Myhrvold C. CRISPR-based surveillance for COVID-19 using genomically-comprehensive machine learning design. *bioRxiv* [Internet] 2020 [cited 2023 Apr 3];2020;2(26); 967026. Available from: www.biorxiv.org/content/10.1101/2020.02.26.967026v2

136. Mortazavi A, Williams BA, Mccue K, Schaeffer L, Wold B. Mapping and quantifying mammalian transcriptomes by RNA-Seq. *nature.com* [Internet] 2008 [cited 2023 Apr 3];5(7); 621. Available from: www.nature.com/articles/nmeth.1226

137. Barh D, Tiwari S, Weener ME, Azevedo V, Góes-Neto A, Gromiha MM, et al. Multi-omics-based identification of SARS-CoV-2 infection biology and candidate drugs against COVID-19. *Computers in Biology and Medicine* [Internet] 2020 [cited 2023 Apr 3];126. Available from: https://pubmed.ncbi.nlm.nih.gov/33131530/

138. Nyholm L, Koziol A, Marcos S, Botnen AB, Aizpurua O, Gopalakrishnan S, et al. Holo-Omics: Integrated host-microbiota multi-omics for basic and applied biological research. *iScience* [Internet] 2020 [cited 2023 Apr 3];23(8). Available from: https://pubmed.ncbi.nlm.nih.gov/32777774/

139. Cocquet J, Chong A, Zhang G, Genomics RV-, 2006 Undefined. Reverse transcriptase template switching and false alternative transcripts. *Elsevier* [Internet] [cited 2023 Apr 3]. Available from: www.sciencedirect.com/science/article/pii/S0888754305003770

Chapter 13

Innovative Therapeutic Efforts for Tackling Future Viral Pandemics

Umme Abiha, Nandan Patel, Neha Goel, and Mohit Sharma

13.1 INTRODUCTION

The transition from hunter-gatherer societies to agrarian communities has facilitated the proliferation of infectious diseases among humans (1). Increased trade between communities has led to greater interactions between humans and animals, resulting in the transmission of zoonotic pathogens. Subsequently, the expansion of cities, extended trade routes, increased travel, and the impact of a growing human population on ecosystems have heightened the emergence and spread of infectious diseases, posing greater risks of outbreaks, epidemics, and even pandemics (2). The terms *endemic, outbreak, epidemic*, and *pandemic* describe the occurrence and spread of a health condition in relation to its expected rate and geographic scope (3). An endemic condition persists at a predictable rate within a population. An outbreak refers to an unexpected increase in the number of individuals affected by a health condition or the occurrence of cases in a new area. An epidemic is an outbreak that spreads across larger geographic regions, while a pandemic is an epidemic that extends globally.

Emerging infections are either newly appearing in a population or spreading to new geographical areas (4). Throughout history, zoonotic transmission of pathogens from animals to humans has been a significant mechanism for the emergence of infections (5). Interactions with animals through activities like hunting, animal farming, trade of animal-based products, wet markets, and the exotic pet trade have substantially increased the likelihood of cross-species transmission of pathogens (6). The process of cross-species transmission involves five stages: (a) the pathogen exclusively infects animals in their natural environment; (b) the pathogen evolves to be transmissible to humans but lacks sustained human-to-human transmission; (c) the pathogen undergoes limited cycles of secondary transmission between humans; (d) the disease exists in animals, but long sequences of secondary human-to-human transmission occur without involving animal hosts; and (e) the disease becomes exclusively present in humans. The risk of zoonotic transmission depends on the animal species harboring the pathogen, the nature of human interaction with that animal, and the frequency of these interactions. Additionally, changes in

DOI: 10.1201/9781003285823-13

land use and climate play crucial roles in the transmission of pathogens from wildlife to humans (7,8). Consequently, surveillance programs are necessary to promptly detect the emergence of pathogens with potential for zoonotic transmission at the animal-human interface.

Climate changes also impact the transmission of various pathogens, such as dengue, chikungunya, Zika, Japanese encephalitis, West Nile virus, and *Borrelia burgdorferi*, by expanding the habitats of disease-carrying vectors like *Aedes albopictus* mosquitoes and ticks (9). The introduction of vector-borne pathogens into non-endemic regions often leads to explosive epidemics. Changes in land use, driven by a growing human population, also influence the distribution of disease-carrying vectors (10). Controlling vector-borne zoonotic pathogens typically requires measures to address the underlying drivers of transmission. Moreover, the spread of several infectious diseases, including tuberculosis, malaria, and cholera, to extended geographic areas is becoming a major health concern for a significant proportion of the population (11). These diseases are spreading more widely due to factors such as the acquisition of drug resistance, increased mosquito vector resistance to insecticides, poor sanitation, land use changes, climate changes, and increased human mobility and travel (12,13).

Cholera outbreaks have been reported in areas affected by natural disasters like earthquakes and floods. To prevent the spread of these pathogens from endemic to non-endemic regions, it is essential to implement surveillance programs.

Certain infectious agents, including *Bacillus anthracis, Yersinia pestis*, and variola virus, have the potential to be used as bioweapons, posing threats to humanity (14). These weapons may involve natural microorganisms or modified microorganisms that are more virulent, easily transmissible, or resistant to treatment. To safeguard the population, governments should establish preparedness plans for biowarfare, bioterrorism, and biocrime (15).

This chapter discusses control and therapeutic measures on current disease management strategies. Infectious diseases continue to pose risks to human health due to their rapid spread through global trade and travel. To address this, global surveillance programs are necessary to detect and identify pathogen spillover from animals to humans, as well as to control waterborne and vector-borne diseases. Moreover, effective preventive and control measures, both pharmaceutical and non-pharmaceutical, are crucial for limiting the dissemination of these infections among the human population.

13.2 GLOBAL RESPONSES BETWEEN THREAT AND HOPE

The COVID-19 pandemic has prompted calls for enhanced global preparedness for future pandemics. These appeals arise from the belief that the current outbreak could have been predicted, prevented, and managed more effectively, leading to fewer social and economic disruptions and loss of lives. Similar calls have been made in the past and have resulted in some positive actions. However, there is a tendency for the world to quickly move on to other crises, leading to a recurring cycle of "panic and neglect." This is concerning because while we cannot predict when or how the next pandemic will occur, we can be certain that it will happen.

There is hope that things will be different this time. The COVID-19 crisis has highlighted the limitations of previous efforts and emphasized the need for a more ambitious and sustained approach to preparedness. It is encouraging to witness widespread calls for increased funding, reform of global health governance, and fresh ideas concerning global public goods. Effective preparedness begins at the national level and involves multiple components. First, strong and resilient healthcare systems, particularly primary care, are necessary to detect disease outbreaks, provide essential care, and facilitate the distribution of vaccines and other medical interventions. Second, surveillance systems and laboratory capacity are crucial for identifying both human and zoonotic disease outbreaks. Third, coordination mechanisms across various sectors are essential for prevention and preparedness. Fourth, legal frameworks and regulatory instruments are required to support outbreak prevention and the deployment of countermeasures. Finally, well-functioning supply chains and adequate stockpiles of essential goods and equipment are needed. These elements of preparedness, along with others, are already established and reflected in the joint external evaluation tool, which originated from the 2005 International Health Regulations (16). Since pathogens do not respect borders, cross-country dimensions of preparedness are also important. Regional and sub-regional institutions play significant roles in harmonizing regulations, establishing reporting standards, sharing information on disease outbreaks, and pooling public health assets such as advanced laboratories and procurement resources.

Many countries, particularly low- and middle-income nations, have long-standing weaknesses in these areas of preparedness, which also translate into regional vulnerabilities. Even countries with stronger preparedness have faced significant challenges during the COVID-19 pandemic, demonstrating the importance of global supply chain resilience, trust-building, cohesive responses, and effective intra-governmental coordination.

At the G20 Summit in November 2020, the World Health Organization (WHO) urged proactive intervention against rising cases, aiming to develop a powerful vaccine against dangerous variants of SARS-CoV-2 and other coronaviruses. Additionally, the need for pan-vaccines or prototype theranostics libraries to counter unknown pathogens was emphasized.

According to the WHO, future pandemics could be caused by the reemergence of viruses, their combinations or mutations, or new pathogens referred to as diseases X. Notably, viruses from various families, such as *Coronaviridae* (SARS, MERS), *Flaviviridae* (West Nile, Zika, yellow fever, dengue virus),

Togaviridae (chikungunya virus), *Arenaviridae* (Lassa fever), and *Filoviridae* (Ebola, Marburg virus) pose significant threats based on their history of epidemics and pandemics in the 21st century. These zoonotic viruses frequently spill over into livestock and other animals, acting as reservoir hosts for transmission to humans. However, predicting the next global outbreak remains challenging, although pre-epidemic forms circulating in reservoirs have been identified (17).

Approximately one year after the outbreak, RNA vaccines were developed rapidly and proved beneficial in mounting a quick response to subsequent waves. However, they are not without limitations, particularly when confronted with rapidly evolving and spreading viruses, as well as issues of vaccine diplomacy and access disparities. Vaccine deployment and the establishment of protective immunity require time. By mid-2022, it was estimated that 70% of the global population would not be fully vaccinated, and the poorest countries would only receive COVID vaccines in 2023. Conversely, conventional anti-infective treatments exhibit inadequate efficacy, limited activity, undesirable side effects, and an increasing rate of drug resistance (18). The discovery of new drugs is a lengthy and costly process, and repurposing existing drugs was considered a faster and more cost-effective alternative, although many repurposed drugs failed, and further studies are needed to evaluate their effectiveness against specific pathogens (19). Furthermore, the long-term efficacy and side effects of experimental drugs against SARS-CoV-2 remain largely unknown.

Considering these factors, the WHO proposed a new international treaty prioritizing pandemic prevention and preparedness, with the aim of detecting and addressing pathogens early and equitably worldwide. Consequently, a shift is needed from a "one bug, one drug" model to more broadly active and adaptable therapeutic approaches (20), which were overlooked previously. Optimizing vaccine platforms and developing next-generation pan-vaccines is crucial. Investment in cutting-edge research is necessary as innovation can save lives. Smart theranostics are needed to outsmart pathogens. It is remarkable that in the 21st century numerous diseases remain incurable, despite advancements in smart technology in various aspects of our lives.

Paradoxically, the ongoing pandemic has propelled the rapid advancement of novel biotechnologies, resulting in significant progress in diverse scientific fields, deeper understanding of respiratory illnesses, and numerous ongoing preclinical and clinical studies. As a consequence, we are entering a new era of revolutionary medicine, with promising and safe smart therapeutic strategies on the horizon.

Given the challenges, the WHO has proposed a new international treaty prioritizing the negotiation of WHO constitution to strengthen global pandemic prevention and preparedness efforts, aiming to detect and address pathogens early and equitably worldwide. Therefore, it is crucial to reassess our approach from a "one bug, one drug" model to more broadly active and adaptable therapeutic strategies (20) that were overlooked previously. We should optimize vaccine platforms to develop next-generation pan-vaccines, invest in cutting-edge advanced research, as innovative approaches save lives. We need smart theranostics to outsmart pathogens before they outsmart us again. It is astonishing that in the 21st century, numerous diseases remain incurable. While we live in smart homes, use smartphones, and build smart cities, prioritizing smart therapies should be our focus to safeguard our lives and health. The ongoing pandemic has accelerated the progress of novel bio-technologies at an unprecedented rate, leading to significant advancements in various scientific fields, including a deeper understanding of respiratory illnesses and a multitude of ongoing (pre)clinical studies. Consequently, we are entering an era of revolutionized medicine, with promising safe and effective smart therapeutic strategies on the horizon.

13.3 SMART THERAPEUTIC INTERVENTIONS

13.3.1 Short Peptides

Peptides, which are RNA-associated pre-proteins, exhibit remarkable intelligence in cellular functions within bio-systems. They serve as effective bio-messengers by regulating intercellular communication through receptor and enzyme activation (21). Over millions of years of evolution, peptides have developed specific molecular recognition capabilities. Originating as catalysts in early life, they have evolved in conjunction with the human body to possess excellent selectivity for specific protein targets. These short peptides combine the advantages of small molecules and biologics. They display unique features such as bio-responsiveness, efficiency, low/no-toxicity, easy design, synthesis, and modification, as well as low production costs and stability in harsh conditions (22). Peptides have the ability to inactivate various pathogens, including those with mutations, by targeting different stages of their life cycle or the host (23). By disrupting protein-protein interactions, which are essential cellular processes and emerging targets, peptides hold significant potential in pharmacy. However, many disease-relevant protein-protein interactions remain unexplored (24). Viral proteins interact with host factors through short peptide interaction motifs in unstructured regions, evolving through mutations to engage novel host factors (25). Understanding these peptide-mediated protein-protein interactions can provide insights into viral tropism and molecular processes within host cells (24).

Peptides offer greater functional and structural diversity than any other molecules, making them highly versatile motifs. Recent advancements in various scientific fields and bio-nanotechnology have addressed the limitations of peptides, such as short half-life and low bioavailability, enabling their extensive use in combating pathogens of any origin (22,26). Consequently, peptides have gained renewed attention, with over 80 peptide-based therapeutics targeting a wide range of diseases, including viral infections, already on the global market. Additionally, more than 800 peptides are currently undergoing (pre)clinical studies (see Table 13.1). Peptides that target protein-protein interactions are particularly interesting due to their potential impact in developing safe and effective drugs (27,28). Notably, three such peptides—nangibotide, reltecimod, and C16G2—are currently in clinical trials for bacterial and viral infections, and the drug discovery process is expected to advance further (29).

Peptide interferons are amyloid-forming peptides that induce the aggregation of proteins, selectively inhibiting viral and bacterial disease processes. These pept-ins disrupt specific proteins, destabilizing pathogens. This technology holds promise for designing broad-spectrum biopharmaceuticals solely based on the peptide's primary sequence, without requiring knowledge of the protein's structure. Notably, this approach has not been applied to any protein to date (30,31).

Peptide aptamers, also known as peptamers or affimers, are fundamental in molecular biology. These synthetic antibodies can serve as substitutes for monoclonal antibodies in various applications. They offer advantages of enhanced stability, ethical considerations, sustainability, cost-effectiveness, and faster production (32).

Peptide nucleic acids (PNAs) carry the encoded information of essential biomolecules, such as proteins and nucleic acids. They hold immense potential in biomedicine due to their ability to target complementary DNA and RNA strands. Unlike DNA/RNA compounds, PNAs lack negative charges in their peptide backbone, resulting in excellent biochemical stability and higher binding energies. This makes PNAs promising scaffolds for innovative solutions in diagnosis, particularly as imaging agents (33).

Short peptides, particularly antimicrobial peptides (AMPs) including host defense peptides, are emerging as powerful broad-spectrum antibiotics. They are effective against drug-resistant microorganisms, demonstrating their significant potential and multifaceted nature. AMPs employ a unique mechanism of action by rapidly killing microbes at low concentrations, exhibiting immunomodulatory abilities and low susceptibility to resistance. Interestingly, many AMPs also display antiviral activity. For instance, Bomidin is effective against various bacteria and enveloped viruses such as SARS-CoV-2, dengue virus, herpes simplex virus, and chikungunya virus. Additionally, a frog-defensin-derived peptide shows promise as a broad-spectrum agent against influenza and diverse variants of SARS-CoV-2. The antiviral and antifungal mechanisms of AMP action have been described elsewhere. Short and modified AMPs offer improved therapeutic efficacy, reduced cytotoxicity, decreased proteolytic digestion, and cost-effective large-scale production. Deep learning-based tools aid in the fast and cost-effective prediction of suitable AMPs. Nanostructured AMPs exhibit enhanced stability, lower toxicity, reduced production costs, prolonged activity, and controlled delivery. Notably, AMPs derived from the secretions of mesenchymal stem cells can mitigate the cytokine storm observed in respiratory diseases. Biopeptides from toxins, post-translationally synthesized peptides, offer new possibilities for severe respiratory syndrome (corona) viruses. Computational tools and databases facilitate the prediction and development of peptide-based drugs, reducing costs and time required before in vitro experiments. Short peptides also serve as specific antigens and adjuvants for modern vaccines. Peptide vaccines have advantages over protein and RNA-based vaccines, including overcoming allergic reactions, autoimmune responses, and generating durable antibody responses. They are safer, relatively inexpensive to produce on a large scale, highly reproducible due to advances in solid-phase peptide synthesis and can be stored for extended periods between outbreaks. Peptide vaccines can avoid immunopathological proinflammatory sequences, off-target antigen loss, and mutant escape. They can incorporate antigens with diverse protective roles or mechanisms, even from different virus proteins. Furthermore, peptide vaccine sequences can be converted into nucleic acids and modified for nucleic (or vector-based) vaccines. Unfortunately, peptide-based vaccines were initially underestimated at the beginning of the pandemic. However, a promising approach to enhance immunogenicity and other effects of peptide-based vaccination is the development of a conjugated self-adjuvating peptide vaccine with an immune agonist.

13.3.2 Antivirals

Researchers from the University of California, Los Angeles (UCLA) have discovered a potential broad-spectrum antiviral treatment capable of targeting multiple families of RNA viruses that pose significant threats for future pandemics. The study, funded by the National Institutes of Health and published in *Cell Reports Medicine*, demonstrated that a stimulator of interferon genes (STING) agonist effectively alleviated the debilitating effects of the chikungunya virus in mice. The COVID-19 pandemic, which has resulted in the loss of nearly 7 million lives worldwide, has exposed humanity's vulnerability to large-scale outbreaks caused by emerging pathogens. Recent epidemics, coupled with global warming and the evolutionary nature of RNA genomes, indicate that arboviruses, such as those transmitted by mosquitoes, are prime candidates for triggering the next pandemic. Among the most prevalent arboviral diseases globally are dengue fever, chikungunya virus, Zika virus, yellow fever, Japanese encephalitis, and West Nile virus. Therefore, it is of utmost importance to discover effective broad-spectrum treatments against these arboviruses due to their demonstrated epidemic potential. In their study, the researchers investigated a collection of innate immune agonists, which target pathogen recognition receptors, and identified several antivirals that inhibited arboviruses to varying extents. However, the most potent and broad-spectrum antiviral agents were cyclic dinucleotide STING agonists.

The STING signaling pathway plays a critical role in generating a strong innate immune response against pathogens. A single dose of cyclic (adenine

Table 13.1 Peptide-Based Therapeutics, in Clinical Use or under Clinical Studies, for Infectious Diseases

S. No.	Peptide-Based Therapeutic	Pathogen	Type/Target	Development Stage	Reference
1	RhACE2-APN01	SARS-CoV-2	hACE2/S protein-hACE2 interaction	Phase II clinical trial (NCT04335136)	34
2	Paxlovid (nirmatrelvir and ritonavir)	SARS-CoV-2	Main protease inhibitor and antiretroviral protease inhibitor and a strong cytochrome P450	Approved by FDA	35
3	Plitidepsin/Solnatide etc. (synthetic peptides)	Influenza virus, respiratory syncytial virus, SARS-CoV (SARS-CoV-2)/SARS-CoV-2	—	III/II	36
4	Canakinumab (anti-inflammatory peptide)	SARS-CoV-2	—	III	37
5	EpiVacCorona peptide antigen-based vaccine	SARS-CoV-2	Peptide design: SARS-CoV-2 proteins conjugated to a carrier protein	Phase III–IV	38
6	B-pVAC-SARS-CoV-2	SARS-CoV-2	Peptide design: SARS-CoV-2–derived multi-peptide vaccine	I/II	39
7	Multimeric-001 (M-001)	Influenza	Composition: influenza hemagglutinin peptides along with standard vaccine	II	40
8	BIPCV/IMX (V512)	Influenza	Composition: influenza viral peptides	I	40
9	Azatanavir (Reyataz)	Human immunodeficiency virus (HIV)	Azapeptide protease inhibitor	—	41

(Continued)

Table 13.1 Peptide-Based Therapeutics, in Clinical Use or under Clinical Studies, for Infectious Diseases (Continued)

S. No.	Peptide-Based Therapeutic	Pathogen	Type/Target	Development Stage	Reference
10	Nisin (polycyclic lantibiotic)	Antibacterial agent (Gram-positive bacteria)	Depolarization of cell membrane	—	42
11	Teicoplatin (Targocid) lipoglyco AMP	Antibacterial agent (Gram-positive bacteria)	Inhibitor of cell wall synthesis	—	41
12	Polymyxin B (poly-Rx) cyclo-lipo AMP	Antibacterial agent (Gram-positive bacteria)	Membrane lysin	—	41
13	Anidulafungin (Eraxis; cyclo-lipo AMP)	Antifungal drug	Inhibitor of the β-(1,3)-D-glucan synthase	—	41
14	Human lactoferrin-derived peptide hLF1–11	Antifungal drug	DNA-binding	—	43

monophosphate–inosine monophosphate), or cAIMP, a STING ligand, triggered a robust host antiviral response that effectively prevented and alleviated the viral arthritis caused by the chikungunya virus in mice. Since individuals affected by the chikungunya virus may suffer from viral arthritis for years or even decades after the initial infection, this treatment approach holds promise. Analysis of transcriptional changes in host cells revealed that cAIMP treatment reversed the detrimental effects of chikungunya-induced dysregulation of cell repair, immune, and metabolic pathways.

In comparison to the West Nile and Zika viruses, chikungunya induces more pronounced transcriptional and chemical imbalances in infected skin cells called fibroblasts at the molecular level. This suggests potential differences in the disease mechanisms among viruses belonging to different families, despite all being transmitted by mosquitoes. Furthermore, the researchers observed that the STING agonists displayed broad-spectrum antiviral activity against both arthropod-borne and respiratory viruses, including SARS-CoV-2 and enterovirus D68, in cell culture models. Apart from their potent antiviral effects against RNA viruses, these STING agonists also show promise in activating immune defenses against cancer. Gustavo Garcia emphasized the next steps of developing these broad-spectrum antivirals in combination with existing treatments and ensuring their availability during future outbreaks of respiratory and arboviral diseases.

Other classes of broad-spectrum antivirals are A-type proanthocyanins (PAC-As), which are natural polyphenols found in plants. They exist as oligomers or polymers of flavan-3-ol monomers like (+)-catechin and (–)-epicatechin, connected by a unique double-A linkage. PAC-As are present in various parts of plants such as leaves, seeds, flowers, bark, and fruits, and are believed to play a protective role against microbial pathogens, insects, and herbivores. When tested individually, PAC-As have demonstrated several biological effects, including antioxidant, antibacterial, immunomodulatory, and antiviral activities. They have shown inhibitory effects on the replication of various human viruses, both enveloped and non-enveloped DNA and RNA viruses. Mechanistic studies have revealed that PAC-As reduce the infectivity of viral particles by interacting with surface proteins essential for viral attachment and entry. Given the significant public health concern posed by viral infections and emerging virus outbreaks, the development of effective broad-spectrum antiviral agents (BSAAs) is urgently needed. PAC-As with antiviral activity are listed in Table 13.2.

13.3.3 Immunotherapy

Immunotherapy plays a crucial role in the early stages of detecting a pandemic. The focus is on rapidly boosting the immunological health of the population. The impact of individual genetic variations and environmental exposure on the implementation of mass immunotherapy is also a

Table 13.2 List of Antiviral Activity of Plant-Derived PAC-As

S. No.	Natural Source	A-Type PACs	Virus and Mechanism of Action	Reference(s)
1	Alpinia zerumbet	Procyanidins*	Influenza A virus, inhibition of attachment, virucidal	44,45
2	Chamaecrista nictitans	Procyanidins	HSV-1 and HSV-2, NA	46
3	Cinnamomum cassia	Dimers, oligomers	HIV-1, interaction with envelope glycoproteins	47
4	Cinnamomum zeylanicum	Trimers, tetramers	HIV-1, inhibition of attachment HCV, inhibition of attachment SARS-CoV, virucidal	48–50
5	Ixora coccinea	Trimers	HIV-1, inhibition of Vpu activity; HCV, NA	51
6	Litchi chinensis	Dimers, trimers	HSV-1 and Coxsackie virus B3, NA	52
7	Pinus maritima	Procyanidins	HIV-1, inhibition of entry and replication HCV, inhibition of replication	53,54
8	Pomelia pinnata	Dimers	HIV-1, inhibition of integrase activity	55
9	Sambucus nigra	Dimers	HIV-1, interaction with envelope glycoproteins	56
10	Theobroma cacao	Dimers	HSV and HIV, NA	57
11	Vaccinium macrocarpon	Monomers, dimers, trimers, tetramers, procyanidins	Nairovirus, inhibition of attachment influenza A and B virus, inhibition of attachment and entry, virucidal	58,59
12	Vaccinium myrtillus	Dimers, trimers	SARS-CoV-2, inhibition of entry and replication	60–62
13	Vitis vinifera	Dimers, trimers, tetramers, procyanidins	Rotavirus, affecting virion integrity HIV-1, inhibition of entry by down-modulation of co-receptors	63,64

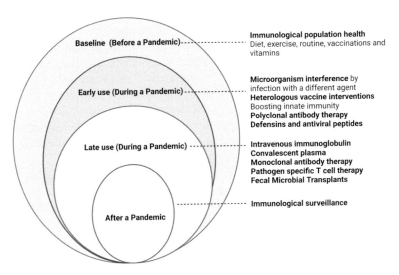

Figure 13.1 Extreme immunotherapies for pandemics timeline.

significant consideration (Figure 13.1). Furthermore, the utilization of mass or individual immunotherapy during a pandemic is explored. Given the emergence of virus variants in the current COVID-19 pandemic, immunotherapy targeting variant escape could serve as a temporary measure while variant vaccines are swiftly developed. Several general principles can be considered, including improving overall population health and immunological health through recommendations such as a healthy diet, increased exercise, and monitoring routine vaccination status (65). However, there are currently no standards for assessing the immunological status of a general population. While the efficacy of vitamins is uncertain, screening and supplementing vitamin D in high-risk groups, particularly the elderly, may enhance general immunological health (66).

Despite extensive efforts to combat the COVID-19 pandemic, ongoing waves of infections and new cases worldwide indicate that factors like politics, fatigue, lack of preparedness, and insufficient resources and infrastructure, particularly in low-income countries, will prolong the pandemic. Interestingly, bystander effects on population-level infectious disease health continue to be discovered, as evidenced by the remarkably low prevalence of influenza A infections in both hemispheres this year. Previous epidemiological studies have demonstrated ecological interactions between infections (67,68). In the context of future pandemics, a radical immunological intervention could involve deliberately infecting the population with a less pathogenic agent that either occupies a niche to prevent target cell availability or is less virulent. The successful implementation of this strategy would require a high percentage of target cells to be infected for an extended period, potentially resulting in a general antiviral effect through the induction of interferons. Although this approach has not been attempted on a large scale, attenuated vaccines operate on a similar principle, and in vitro studies have identified the phenomenon of "viral interference" in target cells (69). Ethical considerations are paramount in providing a less pathogenic agent to protect against a lethal pathogen, particularly as vulnerable individuals may be at risk of disease from the less pathogenic agent. Close monitoring of this approach is necessary to prevent the pathogenic agent from adapting to a broader host range.

Commonly used vaccines can have nonspecific effects, including heterologous effects against SARS-CoV-2 infection. Heterologous vaccine interventions (HVIs) have been proposed as interventions to reduce morbidity and mortality associated with viral pandemics (70,71). The Bacillus Calmette–Guérin vaccine has been ecologically associated with long-term nonspecific benefits and the ability to "boost" innate trained immunity, which could provide protection against various potential pandemic pathogens (72). Routine, age-specific vaccination regimens should be prioritized as one of the initial recommendations when a pandemic is anticipated.

Strengthening the innate component of an individual's immune system is crucial in preventing infection or reducing the severity of a disease in most cases. After infection with a novel pathogen occurs, promptly boosting immunity through interventions such as intravenous immunoglobulin (IVIG) or innate cell therapy could potentially reduce mortality (73). Natural killer (NK) cell responses can be enhanced by administering cytokines like IL-12/IL-18 and IL-15, which have been used in patients with cancer and other conditions. However, cytokine administration carries potential side effects. Bystander actions, such as administering an influenza vaccine, have been shown to non-specifically increase NK cells post-vaccination (74). Another approach involves cellular therapy with "off-the-shelf" NK cells. As knowledge of "innate NK cell memory" and "trained immunity" advances, the elicitation of cross-reactive NK cells against related pathogens may become possible in the future (75–78).

Other innate immune cells, such as γδ, MAIT, and NKT cells, have the potential to be targeted either through ligand engagement or direct cell therapy involving expanded cells. Depending on the specific pathogen causing the pandemic, each of these cell types may play a distinct role (79–83). Therefore, it is crucial for immunologists to be an integral part of the rapid pandemic response scientific teams (84). In certain viral infections, the use of interferon therapy can be considered (85). Interferons not only exhibit antiviral properties but also possess antiproliferative capabilities and can modulate cell differentiation. The choice of interferon class, dosage (high or low), and duration depends on the desired outcome, whether it is antiviral/

antiproliferative activity or immunomodulation. For instance, in the treatment of herpes zoster, administration of high daily doses of interferon to achieve elevated serum levels for several days has been reported (86).

Once a pathogen has been identified and its genetic sequence determined, specific antibodies can be produced either through molecular techniques or by immunizing animals, such as using highly purified F(ab′)2 fragments of horse polyclonal immunoglobulins. Equine polyclonal antibodies containing these fragments are safe, well tolerated, and easily manufactured (87). They can serve as a temporary solution while other antibodies are being developed.

Another rapid option is the administration of IVIG, which is derived from the serum of multiple donors. IVIG exhibits broad antimicrobial activity and has proven effective in reducing bacterial infections when given preventively to patients with chronic lymphocytic leukemia (88) or in cases of human enteroviral encephalitis. However, the timing and method of IVIG administration are crucial. In the later stages of a pandemic, a controlled trial comparing IVIG, convalescent plasma, and a control group would provide definitive results, but conducting such a trial under current clinical guidelines is challenging. While the use of pooled immunoglobulin against SARS-CoV-2 was unsuccessful in clinical trials, combination monoclonal antibody therapy has shown some success in reducing hospitalization rates when administered early (89). However, continuous surveillance for variants and monitoring the effectiveness of monoclonal antibodies is necessary (90). The timing of humoral interventions is critical, as the presence of neutralizing antibodies within the first 14 days of infection is associated with better recovery outcomes in COVID-19 (91). Monoclonal antibody therapy or convalescent plasma is most effective when neutralizing antibodies are present in high quantities and administered as early as possible to suppress viral replication effectively (92,93). This principle likely applies to other pandemic-causing pathogens as well. Adoptive pathogen-specific T cell therapy is also an option, but its usefulness may be limited for acute pathogens, as generating virus-specific T cells takes time (94). Small animal models may become more valuable for assessing virus immunotherapies in the future (95). For a new pandemic pathogen, a strategy worth considering is the administration of an attenuated form of the pathogen, like the use of the attenuated varicella virus vaccine (Zostavax) to prevent symptoms of varicella virus infection (96). This approach requires identifying and isolating the pathogen and inducing mutations through passage to reduce its pathogenicity while still eliciting a protective immune response. However, achieving this within a short time frame is unlikely. It is important to acknowledge the significant potential risks associated with this approach, as well as the potential rewards.

Table 13.3 Emergency Immunotherapy to Combat Pandemics

Adequate diet and exercise
Screen and supplement vitamin deficiencies
Follow vaccination schemata
Heterologous vaccines intervention(s)
Convalescent plasma
Intravenous immunoglobulin
Monoclonal antibodies
Proinflammatory cytokines inhibitors (e.g., interleukin-6/tumor necrosis factor inhibitors)
Complement inhibitors
Interferon therapy
Anti-inflammatory agents (e.g., baricitinib)
Cellular therapy
Mesenchymal stem cell–based therapy
Natural killer cell–based therapy
Unconventional therapies (e.g., fecal microbial transplant)

13.4 POTENTIAL OF ARTIFICIAL INTELLIGENCE TO ACCELERATE PANDEMIC VACCINE DEVELOPMENT

The global community's remarkable effort in developing effective vaccines within a year of the virus's detection, aided by the brilliance, perseverance, and creativity of scientists worldwide, has significantly reduced its fatality rate. Artificial intelligence (AI) has played a crucial role alongside scientists in achieving this feat (Figure 13.3). Moderna, a US company, utilized AI to expedite the development process, enabling them to release an effective COVID-19 vaccine promptly. By employing AI algorithms and robotic automation, Moderna increased its monthly production of mRNA molecules, essential to the vaccine, from around 30 to approximately 1000 (Figure 13.2). AI is also assisting Moderna in mRNA sequence design, contributing to advances in immune medicine. AI technology is driving improvements in various healthcare areas. For instance, a collaborative effort involving Google Health, DeepMind, the UK National Health Service, Northwestern University, and Imperial College London has developed an AI model trained to detect breast cancer from X-ray images. The algorithm demonstrated effectiveness comparable to human radiologists, alleviating healthcare system burdens and enhancing treatment.

The algorithm was crucial in aiding various areas such as genomic taxonomic classification, CRISPR-based detection assays, patient endurance calculation, and identification of potential drug candidates (97,98). Metsky et al. demonstrated the use of machine learning and a CRISPR-based virus detection system for rapid and sensitive screening of SARS-CoV-2 (99).

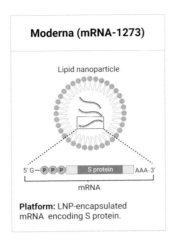

Figure 13.2 Moderna mRNA vaccine.

Additionally, AI can assist in managing critical COVID-19 patients. Rahmatizadeh et al. developed a three-stage model incorporating paraclinical, clinical, and epidemiologic data, personalized medicine, diagnosis, risk stratification, treatment, prognosis, and management. The approach is valuable for patient stratification and timely treatment. Computational tools not only aid in virus detection but also facilitate drug development. Wu et al. analyzed proteins encoded by the SARS-CoV-2 virus, predicting potential targets and candidate drugs, such as 3-chymotrypsin-like protease (3CLpro), spike, RNA-dependent RNA polymerase (RdRp), and papain-like protease (PLpro). In silico studies enable drug repositioning for COVID-19 treatment (100). Furthermore, a deep-learning analysis structure was created for computerized recognition and monitoring of COVID-19 patients using thoracic computed tomography (CT) images. Developing AI-based diagnostic systems can enhance accuracy and speed while reducing healthcare workers' exposure to COVID-19 patients.

In the case of malaria, AI-powered handheld lab-on-a-chip molecular diagnostics systems, developed by the Digital Diagnostics for Africa Network in partnership with MinoHealth AI Labs and Imperial College London's Global Development Hub, offer a revolutionary approach to detecting the disease in remote parts of Africa (101). This technology has the potential to facilitate universal health coverage and contribute to achieving UN Sustainable Development Goal 3. Despite the benefits of AI in medicine, its integration into society and innovation faces challenges. Concerns over facial recognition, automated decision-making, and COVID-related tracking have prompted caution and suspicion. To build trust and inclusivity, it is crucial to employ AI in transparent and internationally coordinated ways, mirroring the collaborative approach seen during the pandemic (102).

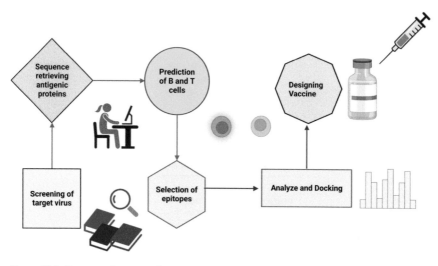

Figure 13.3 Process of vaccine discovery by AI/ML method.

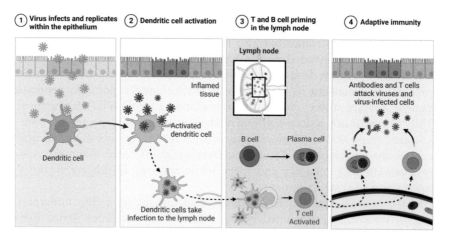

Figure 13.4 Schematic representation of adaptive immunity targeting pathogen for killing.

13.5 SYSTEMS BIOLOGY TO ASSESS THE POTENTIAL OF THERAPEUTIC AGENTS

Next-generation technologies have enabled the application of systems biology in assessing microbial virulence and associated pathogens (103,104). To address the emergence of novel pathogens resulting from mutations, gene transfer, and outbreaks, it is crucial to update databases with relevant information and risk assessments. This presents both a challenge and an opportunity for multiomics experts to collect and organize new datasets or update existing databases. The availability of this information will aid clinical and medical researchers in planning research methodologies for therapeutic and drug development. For example, the novel coronavirus has become a global nuisance, causing a significant number of infections and deaths. Risk assessment and exposure characterization are essential for understanding the virus and developing potential therapeutics.

Systems biology, including multiomics, databases, and in silico studies, plays a major role in this field. In silico approaches rely on computational design and protein interaction analysis, often utilizing database mining. Researching and gathering information are prerequisites for database development and practical methodology planning. Epigenomics, the study of heritable phenotypic changes, may contribute to understanding mortality rates among different population groups (105,106). Epigenetic mechanisms, such as modifications on lysine acetylation, methylation, phosphorylation, and ubiquitination, can influence genetic and phenotypic changes. The epigenetic control of ACE2 in lung cells may play a role in the susceptibility to coronavirus infection. Further exploration of epigenetics is needed to prevent and treat COVID-19.

DNA methylation may also affect the vulnerability of viral genetic material. Genomics, proteomics, and metabolomics correspond to DNA, RNA, proteins, and metabolites, respectively. Other systems biology approaches like immunomics, host lipid omics, public health omics, and quantitative dynamic omics can contribute to the development of effective drugs against COVID-19.

The interaction between proteins and receptors relies on specific receptor-binding domains (105). These domains recognize and bind to interactive sequences on receptors through non-covalent bonds in the human system. However, SARS-CoV-2 primarily targets the mucous membranes as its initial attachment site, then binds to its receptor, ACE2, enabling the viruses to enter host cells (106). Once inside the cell, viral replication occurs with the assistance of replication proteins and subsequent multiplication steps (Figure 13.5). During replication and amplification, SARS-CoV-2 releases proteases, resulting in the generation of reactive oxygen species (ROS) in the host cell (106,107). These ROS are toxic to the cell and its surroundings. In response, the immune system is activated, and neutrophils and monocytes migrate to the lungs, particularly the alveoli, where the virus has reached after attaching to nasal, oral, or mucous surfaces. Neutrophils, basophils, and monocytes are myeloid progenitor cells and part of the innate immunity. They are upregulated upon encountering foreign particles, including viruses (108). The release of cytokines is initiated through cell signaling in both the adaptive and innate immune systems. Macrophages, which differentiate from monocytes, undergo metabolic changes in response to toxicity and foreign particles, leading to the release of inflammatory cytokines.

Transmission Mechanism of SARS-CoV-2

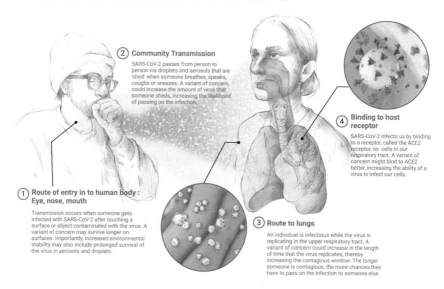

Figure 13.5 Transmission mechanism of SARS-CoV-2.

Table 13.4 In Silico Tools Used in (Protein-Protein/Protein-Ligand Study for Drug Target and Protein Target Inhibition) Studies

S. No.	In Silico Tool	Function	Reference
1	Chemdraw	Ligand structure analysis	109
2	SWISS-MODEL	Protein structure homology modeling	110
3	ERRAT	Protein structure validation	111
4	ProSA	Protein structure validation	112
5	ProQ	Protein structure validation	113
6	RAMPAGE	Protein structure validation	114
7	UCSF Chimera	Protein analysis	115
8	PyMOL	Protein analysis	116
9	AutoDock	Molecular docking	117
10	Schrodinger	Molecular docking	118

These cytokines affect molecular signaling in target cells and neighboring cells. Disruptions in cellular signaling caused by ROS stress or inflammatory cytokines from SARS-CoV-2 can result in mucus accumulation in the lungs, difficulty breathing, abnormal blood pressure, and other symptoms. To enhance our understanding of the host-virus interaction, a small-scale study utilizing computational biology and in silico tools has been designed.

13.6 SECONDARY INFECTION PREPAREDNESS

Secondary bacterial infections frequently occur in conjunction with or following respiratory viral infections. These viral infections impair the respiratory airways and compromise both the innate and acquired immune responses, creating a favorable environment for bacterial proliferation, adherence, and invasion into healthy areas of the respiratory tract. Investigating the molecular mechanisms underlying viral-induced secondary bacterial infections offers an opportunity to develop innovative and effective therapeutic strategies for disease prevention. Respiratory viruses often promote the growth of opportunistic bacterial pathogens, thereby exacerbating secondary bacterial infections. Viral infections cause histological and functional damage to the respiratory airway, characterized by cell loss, goblet cell hyperplasia, altered mucus secretion, reduced ciliary beat frequency, impaired mucociliary clearance function, and compromised oxygen exchange (119,120). These effects are associated with various molecular mechanisms that contribute to the increased susceptibility of the respiratory tract to bacterial infection. The majority of documented interactions between bacteria and viruses involve viral infections in the gastrointestinal tract. Commensal bacteria in the body's system serve as the first line of defense by outcompeting disease-causing bacteria and limiting their access to tissues. However, some viruses have evolved to exploit this contact with commensal bacteria, facilitating the development of disease. In laboratory conditions, viruses can directly bind to target cells and replicate. However, in the gastrointestinal tract, this strategy faces challenges due to the presence of a large number of bacteria that compete for receptor binding sites, reducing the likelihood of pathogenic bacterial growth or virus attachment. Other factors like mucus or enzymatic secretions can also hinder or assist the infection process. To overcome these obstacles, instead of competing for host cell binding sites, some viruses can utilize bacterial molecules to enhance their association with more accessible host cells, initiating infection. This strategy can be employed by viruses that target cell types other than epithelial cells or in addition to epithelial cells.

Although the nature of the interaction remains the same, viruses that interact with bacteria may have additional benefits beyond disease progression. Studies have shown that association with fecal microbiota increases the environmental fitness and stability of poliovirus. Exposure to bacteria or their polysaccharides decreases the efficacy of virus inactivation by heat and bleach, potentially aiding viral survival in the environment. This observation is supported by the higher susceptibility to inactivation with heat observed in a poliovirus mutant that does not bind to bacterial polysaccharides as efficiently. Additionally, bacterial polysaccharide components enhance the binding of wild-type poliovirus to its host cells, expressing its receptor. In summary, gastrointestinal microbiota not only increase poliovirus infectivity but also promote virus transfer to the next host. Thus, direct viral interaction with bacteria plays a role in viral pathogenesis and is an emerging area of study in microbiology.

Viral infections can also benefit bacterial species, as the disease state induced by the virus allows normally harmless bacteria to become opportunistic pathogens. Under normal healthy conditions, competition between microbes limits the invasion of pathogens by saturating colonization sites, boosting the immune response, and producing antimicrobials. However, when microbial populations are disrupted, previously inaccessible niches become available to invading pathogens, compromising the microenvironment where native microbiota outcompeted disease-causing counterparts. The mechanisms through which viruses aid bacterial pathogenesis include upregulation of cellular receptors, disruption of epithelial layers, displacement of commensal bacteria, and suppression of the immune system. Influenza viruses and pathogenic bacteria such as *Streptococcus pneumoniae*, *Staphylococcus aureus*, and *Haemophilus influenzae* are well-studied examples of these interactions within the human body. Influenza viruses damage the host epithelium and provide potential binding sites for bacteria through neuraminidase cleavage of sialic acid, upregulation of bacterial host receptors, and host regeneration of common bacterial receptors. This pattern of host damage is common in upper respiratory tract viruses and bacteria. The interactions between influenza viruses and bacteria exemplify the multiple complex ways in which viral infection can indirectly assist bacterial infection. Refer to Table 13.5 for secondary bacterial diseases known to be associated with specific viral infections.

Bacterial and fungal infections commonly complicate viral pneumonia, particularly in critically ill patients, leading to increased mortality and intensive care requirements. Influenza patients often experience bacterial coinfection, occurring in approximately 0.5% of healthy young individuals and at least 2.5% of older individuals. During the H1N1 pandemic in 2009, approximately 25% of patients had bacterial or fungal infections. Limited data is available regarding bacterial and fungal infections in viral pneumonia caused by coronavirus. A cohort study during the 2003 SARS outbreak found that 70.6% of critical SARS patients who underwent invasive procedures had secondary lower respiratory tract infections. The pathogens causing these infections varied, with negative bacilli being the most common. Invasive pulmonary aspergillosis was also a common complication of influenza. The investigation and reporting of bacterial and fungal infections in COVID-19 patients have been insufficient. Only a small number of articles have reported secondary infections, often lacking detailed information on the pathogens involved. Antibiotic use rates in these studies were much higher than the reported incidence of secondary infections. Furthermore, most published papers analyzing prognosis did not include bacterial or fungal infections as complications. Existing infection control protocols primarily focus

Table 13.5 List of Secondary Bacterial Infections and Their Immune Responses during Viral Infection

S. No.	Viral Infection	Secondary Bacterial Disease	Bacterial Name	Immune Response
1	Influenza	Pneumonia, otitis media, sinusitis, meningitis	Streptococcus pneumoniae, Staphylococcus aureus, Streptococcus pyrogenes, Moraxella catarrhalis, Neisseria meningitidis	Loss of ciliary function, produce INF, mucus response. Destroy phagocytic cell, produce cytokines including TNF-α, IL-6, and pro-IL-1β
2	Coronavirus	Pneumonia	Haemophilus influenzae	Activate the innate immune response or induce costimulatory signals for adaptive immunity
3	Adenovirus	Pneumonia	S. pneumoniae, H. influenzae, S. aureus, M. catarrhalis	Produce cytokines, loss of ciliary function, produce INF, mucus response
4	Measles virus	Otitis media, pneumonia, tracheobronchitis	S. pneumoniae, H. influenzae, S. aureus	Activate the innate immune response, produce cytokines including TNF-α, IL-6, and pro-IL-1β, mucus response
5	Human rhinovirus	Pneumonia, otitis media, sinusitis	S. pneumoniae, H. influenzae, S. aureus	Alters in pulmonary immune response, produce cytokines, inflammatory cell recruitment
6	Parainfluenza virus	Pneumonia	S. pneumoniae	Destroy phagocytic cell, produce cytokines including TNF-α, IL-6, and pro-IL-1β
7	Respiratory tract viral infection	Pneumonia, lungs problem	S. pneumoniae, S. aureus, H. influenzae, Pseudomonas aeruginosa	Mucus response, IFN secretion, epithelial cell death, cytokine release, loss of ciliary function
8	Urinary tract viral infection, BK virus, herpesvirus	Urinary tract infection	Escherichia coli, Klebsiella pneumoniae, P. aeruginosa	Chemokines and cytokines such as CXCL8, CCL2, interleukins (IL-6, IL-8, IL-10, IL-17A), and granulocyte colony-stimulating factor (G-CSF) are released

on preventing the transmission of SARS-CoV-2, neglecting the prevention of secondary infections. However, secondary infections have been found in 50% of non-surviving COVID-19 patients. Diagnostic challenges arise from the difficulty of differentiating bacterial or fungal infections from existing viral pneumonia based on clinical and radiological appearance. Microbiological examinations, including sputum culture, can aid in diagnosis but pose risks to laboratory personnel due to the transmission of the virus through aerosols and direct contact.

Standardized diagnostic processes for secondary infections and personal protection guidelines for healthcare workers processing bacterial and fungal cultures are lacking. Inadequate laboratory biosafety conditions and shortages of personal protective equipment (PPE) hinder routine microbiological examinations in COVID-19 patients. To address these issues, it is crucial to strengthen the investigation of secondary infections in COVID-19 patients while ensuring the safety of laboratory staff. Practical diagnostic processes for bacterial and fungal infections should be developed, along with clear guidelines for biosafety and PPE requirements. Qualified medical institutions should be encouraged to conduct necessary microbiological examinations. These efforts will provide valuable clinical data for evidence-based treatment, prevention, and control of infection complications, ultimately reducing the mortality associated with COVID-19. Since December 2019, the global COVID-19 pandemic has impacted over 539 million individuals, resulting in more than 6.3 million deaths. The causative agent, severe acute respiratory syndrome coronavirus 2 (SARS-CoV-2), is a novel coronavirus that primarily affects the respiratory system. It shares genetic similarities with the SARS-CoV virus from 2002 and the MERS-CoV virus from 2012. While most infected individuals experience mild to moderate symptoms, those with underlying health conditions may require hospitalization and intensive care, making them susceptible to secondary infections.

During hospitalization, COVID-19 patients may experience changes in their clinical condition due to various factors, including blood clot–related diseases, adverse effects of medications, and vascular complications. Therefore, it is crucial for clinicians to consider common infections in the diagnostic process and be aware of their characteristic clinical and imaging features. Failure to recognize these secondary infections could result in missed opportunities to diagnose preventable diseases. Any unexpected alterations in the radiographic presentation of COVID-19, such as the appearance of new areas of consolidation or cavities, should prompt further investigation to rule out additional opportunistic infections. While the pulmonary consequences of COVID-19 have been extensively studied, there remains a lack of comprehensive documentation on the manifestations and radiological findings of secondary infections. The imaging features of COVID-19 often lack specificity or deviate from the typical patterns, which poses challenges in accurately diagnosing and effectively treating co-infections in COVID-19 patients.

A classic example is that of mucormycosis (black fungus) that is caused by Mucorales. It is a severe opportunistic fungal infection characterized by tissue necrosis due to angioinvasion and subsequent thrombosis. The infection mainly affects immunocompromised individuals and can manifest in various clinical forms, including pulmonary, sinusitis, gastrointestinal, cutaneous, and disseminated disease. Unlike invasive aspergillosis, the prognosis of mucormycosis has not significantly improved in recent years due to challenges in isolating the causative fungi, lack of reliable diagnostic biomarkers, and limited efficacy of antifungal agents against Mucora. Prompt diagnosis and a high level of suspicion are crucial for early intervention, as mucormycosis carries a high mortality rate. In hospitalized COVID-19 patients, there have been reports of co-infection with mucormycosis (CAM), often leading to fatal outcomes. While these patients may have multiple risk factors predisposing them to the infection, the immune alteration caused by SARS-CoV-2 itself, along with immunosuppressant and high-dose steroid use during COVID-19 treatment, further increases susceptibility to fungal infections. Garg et al. described a case of probable pulmonary mycosis in a patient with diabetes and end-stage renal disease initially diagnosed with COVID-19 pneumonia. The patient showed improvement and resolution of COVID-19 symptoms after 14 days of treatment with intravenous dexamethasone, remdesivir, and supportive therapies. However, 3 days later, the patient developed new symptoms such as cough with expectoration and burning micturition. Imaging revealed a pulmonary cavity with intracavitary contents in the right upper lobe, confirmed by subsequent CT scan, and associated with minimal pleural effusion on the right side. Rhizopus microspores were isolated from the sputum, and the patient received treatment with liposomal amphotericin B for probable pulmonary mucormycosis. Symptoms improved, and the patient was discharged 54 days after initial hospitalization. Similar case reports of COVID-19-associated pulmonary mucormycosis have also been documented.

Similar to previous respiratory viral infections such as the 1918 influenza outbreak and the 2009 H1N1 pandemic, the current pandemic caused by SARS-CoV-2 has been linked to secondary bacterial infections, leading to poor outcomes and fatalities. While recent studies have reported a low rate of secondary bacterial and fungal infections in COVID-19 patients, it is important to acknowledge the limited knowledge we currently have about this novel coronavirus. Some recent cohort studies have found SARS-CoV-2 associated infections in patients with tuberculosis (TB), but more research is needed to understand the impact of latent or active TB on COVID-19 severity and progression. Another cohort study suggested that pre-existing chronic obstructive pulmonary disease or secondary bacterial pneumonia may contribute to the most severe outcomes in COVID-19 cases.

SARS-CoV-2 shares similarities with the SARS-CoV virus, which has been shown to regulate the expression of immune function-related genes in human monocytes. This downregulation and differential regulation of immune genes may create an environment conducive to secondary bacterial infections. Previous studies on SARS patients have confirmed the presence of various bacteria in human samples, including *Escherichia coli, Klebsiella pneumoniae, Pseudomonas aeruginosa*, and methicillin-resistant *S. aureus* (MRSA), suggesting the role of secondary bacterial infections in viral disease progression. A recent study on pregnant women confirmed SARS-CoV-2 infection and investigated its clinical features and neonatal outcomes. The study suggested that timely use of antibiotics could prevent secondary bacterial infections associated with COVID-19, thereby reducing mortality and complications during pregnancy.

SARS-CoV-2 RNA has also been detected in stool samples of COVID-19 patients, indicating the possibility of gastrointestinal infection and a potential fecal-oral route of disease transmission. The high expression levels of ACE2 mRNA in the gastrointestinal system provide indirect evidence of SARS-CoV-2 infection in this area. Preliminary studies have suggested the involvement of the gastrointestinal system in SARS-CoV-2 infection and its interaction with the diverse gut microbiome, which may contribute to immune suppression and secondary bacterial infections. A recent review discussed secondary bacterial and fungal infections in coronavirus patients and provided guidelines for the use of antimicrobials in SARS-CoV, SARS-CoV-2, and related viral infections. However, further clinical studies are necessary to shed light on the relationship between SARS-CoV-2 and secondary bacterial infections.

13.7 IDENTIFYING PATHOGENS FOR FUTURE PANDEMIC POTENTIAL

The WHO is launching a global scientific initiative to update the list of priority pathogens, which are disease-causing agents that can lead to outbreaks or pandemics. The purpose of this initiative is to provide guidance for global investment, research, and development, particularly in the areas of vaccines, tests, and treatments. Over 300 scientists convened at WHO on November 18th 2022 to discuss on a list of priority pathogens that harbor the potential for future healthcare catastrophes. These experts will examine evidence related to more than 25 virus families, bacteria, and an unidentified pathogen known as "Disease X," which represents a potential unknown pathogen that could cause a significant international epidemic. The scientists will propose a list of priority pathogens that require further research and investment. The selection process will consider scientific and public health criteria, as well as criteria associated with socioeconomic impact, accessibility, and equity. The initial publication of the list occurred in 2017, and the most recent prioritization exercise was conducted in 2018. The current list includes COVID-19, Crimean-Congo hemorrhagic fever, Ebola virus disease and Marburg virus disease, Lassa fever, Middle East respiratory syndrome

(MERS) and severe acute respiratory syndrome (SARS), Nipah and henipa-viral diseases, Rift Valley fever, Zika, and Disease X.

Dr. Michael Ryan, executive director of WHO's Health Emergencies Programme, emphasized the importance of targeting priority pathogens and virus families for research and development to enable a rapid and effective response to epidemics and pandemics. Dr. Ryan highlighted that the timely development of safe and effective vaccines would not have been possible without substantial investments in research and development prior to the COVID-19 pandemic. For the identified priority pathogens, the WHO R&D blueprint for epidemics provides R&D roadmaps that outline knowledge gaps and research priorities. Additionally, target product profiles are developed to guide developers in creating vaccines, treatments, and diagnostic tests according to desired specifications. Efforts are also made to map, compile, and facilitate clinical trials to advance the development of these tools. Complementary actions, including strengthening regulatory and ethical oversight, are also taken into account.

Dr. Soumya Swaminathan, WHO Chief Scientist, emphasized that the list of priority pathogens serves as a reference point for the research community to direct their efforts in managing future threats. The list is developed in collaboration with field experts and represents the agreed direction for global research communities to invest their resources in developing tests, treatments, and vaccines. The WHO expresses gratitude to donors, such as the US government, partners, and scientists working with the organization for their contributions to this endeavor.

13.8 HOW CAN FUTURE PANDEMICS BE PREVENTED?

The initial spread of plague and cholera pandemics occurred along trade and military routes. Subsequently, the movement of human populations through rail, ship, and air travel facilitated the geographic dissemination of pathogens. Nowadays, the globalization of travel and trade, including the exchange of animals and animal-based foods, further amplifies the transmission of infectious diseases and accelerates their worldwide spread. Changes in land use and urbanization, driven by agricultural and residential expansion, alter the habitats of pathogens, hosts, and disease vectors, thereby impacting the transmission dynamics of infections to humans. Climate change also influences the distribution of disease vectors, hosts, and microorganisms, potentially increasing the spread of pathogens. Increased human-animal interactions through activities such as breeding, hunting, wet markets, and exotic pet trade increase the risk of spillover of zoonotic pathogens. As a result of human activities and their impact on the environment, the spread of infectious diseases is expected to rise, leading to more frequent epidemics and pandemics that pose new challenges for public health.

To combat the transmission of water-borne pathogens like *Vibrio cholerae*, the WHO has initiated a water, sanitation, and hygiene (WASH) program in developing countries (121). This program emphasizes the provision of safe and clean water sources, effective sanitation infrastructure, and promotion of proper hygiene practices. As reported by the United Nations Children's Fund (UNICEF) and WHO (2019), the WASH program has increased access to safe drinking water, safe sanitation services, and basic handwashing facilities with soap and water in approximately 71%, 45%, and 60% of the global population, respectively. Vector control plays a crucial role in managing vector-borne diseases such as malaria, dengue virus, chikungunya virus, and Zika virus (122). These control measures target either the immature stages of vectors through the use of predator species, larvicides, or habitat modification, or the adult vectors through the application of nets, topical repellents, insecticides, and spraying. Additionally, new vector control methods are being developed, including genetic manipulation of mosquitoes (123), bacterial infection of vectors (e.g., Wolbachia) (124), and eave tubes with insecticide-laden electrostatic netting (125). However, there is still a need for the development of novel vector control tools.

Implementing global surveillance programs for the early detection of pathogen spillover from animals to humans is of utmost importance. The One Health concept promotes optimal health for humans, animals, and the environment. The environmental impacts resulting from land use, urbanization, and climate change increase the risk of zoonotic spillover, underscoring the importance of an integrated One Health approach to zoonosis surveillance. Such interdisciplinary efforts are employed to establish surveillance programs for the prevention and control of emerging and re-emerging infections in developing countries, which are disproportionately affected by zoonoses.

Viral zoonoses pose a significant public health threat, considering that recent pandemics have originated from viruses. Therefore, it is crucial to assess the risk of viruses transmitting between different species and infecting humans. The potential for zoonotic transmission of mammalian viruses can be predicted by the variety of viruses associated with a particular animal species. Research has demonstrated that the likelihood of cross-species transmission of mammalian viruses increases as the hosts and humans become more phylogenetically similar. Additionally, viruses that infect a wide range of hosts across different phylogenetic groups are more likely to be zoonotic. By characterizing the virus diversity in key wildlife species, we can reduce the time between outbreak detection and response. To accomplish this, the global virome project was launched, aiming to detect and identify viral threats to human health, determine the host ranges of viruses, identify behaviors that promote spillover events, establish a global surveillance network, and identify markers for transmission and pathogenicity of high-risk viruses. Moreover, viral metagenomic next-generation sequencing analysis of nasal and throat swabs from individuals at risk of zoonotic infections has been employed to expand the detection of new viruses and characterize the respiratory virome of humans exposed to animals.

13.9 CONCLUSION

The timing and causative agent of the next pandemic cannot be predicted. Therefore, pandemic preparedness plans stress the implementation of non-pharmaceutical interventions as the initial approach to control person-to-person transmission of the pathogen. These interventions should effectively limit the spread of infection while minimizing societal and economic disruptions. However, there is a risk of resurgence once these interventions are lifted. To enhance response effectiveness, it is essential to establish rapid testing, contact tracing, and isolation of infected individuals. Additionally, the development of pharmaceutical interventions such as rapid point-of-care diagnostic tests, biomarkers for disease stratification, broad spectrum antimicrobials/antivirals through in silico drug repurposing, drugs targeting host cells, and new platforms for accelerated vaccine development and production would significantly improve the global response to pandemics.

In the 21st century, infectious outbreaks are unavoidable, although pandemics can be prevented. Science possesses the means to save numerous lives. This chapter explores the challenges posed by pandemics and highlights the tremendous potential of short peptides, both synthetic and nature inspired, which are cost-effective and easily developed in terms of time and technology. These peptides offer intelligent therapeutic strategies that can stimulate further research in this field. The emergence of smart peptide-based approaches is revitalizing the arsenal against infectious diseases, revolutionizing medicine, and presenting boundless therapeutic opportunities.

However, to ensure the successful realization of ambitious scientific goals in smart therapies, it is essential to secure strong funding support and adopt more holistic anti-pandemic strategies. Considerable efforts are still required to establish more advanced global health networks and foster trust in pharmaceutical companies to address vaccine hesitancy. Moreover, it is crucial to implement sustainable development mechanisms and shift towards a harmonious approach to life, protecting the planet's biodiversity and eliminating warfare as proactive "vaccines." The safety of individuals globally relies on the safety of everyone. Therefore, it is imperative to continue global cooperation to swiftly advance the development of cutting-edge smart therapies, address pertinent questions to enhance therapeutic options, and ultimately prevent and treat viral and bacterial infectious diseases with unprecedented speed.

REFERENCES

1. Dobson AP, Carper ER. Infectious diseases and human population history: Throughout history the establishment of disease has been a side effect of the growth of civilization. *Bioscience* 1996;46; 115–126.
2. Lindahl JF, Grace D. The consequences of human actions on risks for infectious diseases: A review. *Infection Ecology & Epidemiology* 2015;5; 30048. Available from: https://doi.org/10.3402/iee.v5.30048
3. Grennan D. What is a pandemic? *JAMA* 2019;321; 910. Available from: https://doi.org/10.1001/jama.2019.0700
4. Morens DM, Folkers GK, Fauci AS. The challenge of emerging and re-emerging infectious diseases. *Nature* 2004;430; 242–249. Available from: https://doi.org/10.1038/nature02759
5. Wolfe ND, Dunavan CP, Diamond J. Origins of major human infectious diseases. *Nature* 2007;447; 279–283. Available from: https://doi.org/10.1038/nature05775
6. Bengis RG, Leighton F A, Fischer JR, Artois M, MornerT, Tate CM. The role of wildlife in emerging and re-emerging zoonoses. *Scientific and Technical Review* 2004;23; 497–511.
7. Wolfe ND, Dunavan CP, Diamond J. Origins of major human infectious diseases. *Nature* 2007;447; 279–283. Available from: https://doi.org/10.1038/nature05775
8. El-Sayed A, Kamel M. Climatic changes and their role in emergence and re-emergence of diseases. *Environmental Science and Pollution Research* 2020;27; 22336–22352. Available from: https://doi.org/10.1007/s11356-020-08896-w
9. White RJ, Razgour O. Emerging zoonotic diseases originating in mammals: A systematic review of effects of anthropogenic land-use change. *Mammal Review* 2020. Available from: https://doi.org/10.1111/mam.12201
10. Caminade C, McIntyre KM, Jones AE. Impact of recent and future climate change on vector-borne diseases. *Annals of the New York Academy of Sciences* 2019;1436; 157–173. Available from: https://doi.org/10.1111/nyas.13950
11. Kilpatrick AM, Randolph SE. Drivers, dynamics, and control of emerging vector-borne zoonotic diseases. *Lancet* 2012;380; 1946–1955. Available from: https://doi.org/10.1016/S0140-6736(12)61151-9
12. Morens DM, Folkers GK, Fauci AS. The challenge of emerging and re-emerging infectious diseases. *Nature* 2004;430; 242–249. Available from: https://doi.org/10.1038/nature02759
13. Cutler SJ, Fooks AR, van der Poel WH. Public health threat of new, reemerging, and neglected zoonoses in the industrialized world. *Emerging Infectious Diseases* 2010;16; 1–7. Available from: https://doi.org/10.3201/eid1601.081467
14. Oliveira M, Mason-Buck G, Ballard D, Branicki W, Amorim A. Biowarfare, bioterrorism and biocrime: A historical overview on microbial harmful applications. *Forensic Science International* 2020;314; 110366. Available from: https://doi.org/10.1016/j.forsciint.2020.110366
15. Narayanan N, Lacy CR, Cruz JE, Nahass M, Karp J, Barone JA, et al. Disaster preparedness: Biological threats and treatment options. *Pharmacotherapy* 2018;38; 217–234. Available from: https://doi.org/10.1002/phar.2068
16. Available from: www.who.int/publications/i/item/9789240051980
17. Meganck RM, Baric RS. Developing therapeutic approaches for twenty-first-century emerging infectious viral diseases. *Nature Medicine* 2021;27; 401–410. Available from: https://doi.org/10.1038/s41591-021-01282-0
18. Pour PM, Fakhri S, Asgary S, Farzaei MH, Echeverría J. The signaling pathways, and therapeutic targets of antiviral agents: Focusing on the antiviral approaches and clinical perspectives of anthocyanins in the management of viral diseases. *Frontiers in Pharmacology* 2019;10; 1207–1230. Available from: https://doi.org/10.3389/fphar.2019.01207
19. Hanisch M, Rake B. Repurposing without purpose? Early innovation responses to the COVID-19 crisis: Evidence from clinical trials. *R&D Management* 2021;51; 393–409. Available from: https://doi.org/10.1111/radm.12461
20. Dolgin E. The race for antiviral drugs to beat COVID—And the next pandemic. *Nature* 2021;592; 340–343. Available from: https://doi.org/10.1038/d41586-021-00958-4
21. Khavinson VK, Popovich IG, Linkova NS., Mironova ES, Ilina AR. Peptide regulation of gene expression: A systematic review. *Molecules* 2021;26; 7053. Available from: https://doi.org/10.3390/molecules26227053

22. Apostolopoulos V, Bojarska J, Chai TT, Elnagdy S, Kaczmarek K, Matsoukas J, et al. A global review on short peptides: Frontiers and perspectives. *Molecules* 2021;26; 430–475. Available from: https://doi.org/10.3390/molecules26020430

23. Lee YC, Shirkey JD, Park J, Bisht K, Cowan AJ. An overview of antiviral peptides and rational biodesign considerations. *BioDesign Research* 2022; 1–19. Available from: https://doi.org/10.34133/2022/9898241

24. Cabri W, Cantelmi P, Corbisiero D, Fantoni T, Ferrazzano L, Martelli G, et al. Therapeutic peptides targeting PPI in clinical development: Overview, mechanism of action and perspectives. *Frontiers in Molecular Biosciences* 2021;8; 697586–697607. Available from: https://doi.org/10.3389/fmolb.2021.697586

25. Kruse T, Benz C, Garvanska DH, Lindqvist R, Mihalic F, Coscia F, et al. Large scale discovery of coronavirus-host factor protein interaction motifs reveals SARS-CoV-2 specific mechanisms and vulnerabilities. *Nature Communications* 2021;12; 6761–6774. Available from: https://doi.org/10.1038/s41467–021–26498-z

26. Muttenthaler M, Adams DJ., Alewood PF. Trends in peptide drug discovery. *Nature Reviews Drug Discovery* 2021;20; 309–325. Available from: https://doi.org/10.1038/s41573-020-00135-8

27. Parra ALC, Bazerra LP, Shawar DE, Neto NAS, Mesquita FP, da Silva GO, et al. Synthetic antiviral peptides: A new way to develop targeted antiviral drugs. *Future Virology* 2022; 14. Available from: https://doi.org/10.2217/fvl-2021–0308

28. Wang L, Wang N, Zhang W, Cheng X, Yan Z, Shao G, et al. Therapeutic peptides: current applications and future directions. *Signal Transduction and Targeted Therapy* 2022;7; 48. Available from: https://doi.org/10.1038/s41392-022-00904-4

29. Cabri W, Cantelmi P, Corbisiero D, Fantoni T, Ferrazzano L, Martelli G, et al. Therapeutic peptides targeting PPI in clinical development: Overview, mechanism of action and perspectives. *Frontiers in Molecular Biosciences* 2021;8; 697586–697607. Available from: https://doi.org/10.3389/fmolb.2021.697586

30. Michiels E, Roose K, Gallardo R, Khodaparast L, Khodaparast L, Kant R, et al. Reverse engineering synthetic antiviral amyloids. *Nature Communications* 2020;11; 2832–2845. Available from: https://doi.org/10.1038/s41467-020-16721-8

31. Wu G, Khodaparast L, Khodaparast L, De Vleeschouwer MD, Hiusmans J, Houben B, et al. Investigating the mechanism of action of aggregation-inducing antimicrobial peptins. *Cell Chemical Biology* 2021;28; 524–536;e4. Available from: https://doi.org/10.1016/j.chembiol.2020.12.008

32. Kruger A, Santos AP, Sa V, Ulrich H, Wrenger C. Aptamer applications in emerging viral diseases. *Pharmaceuticals* 2021;14(7); 622. Available from: https://doi.org/10.3390/ph14070622

33. Exner RM, Paisey SJ, Redman JE, Pascu SI. Explorations into peptide nucleic acid contrast agents as emerging scaffolds for breakthrough solutions in medical imaging and diagnosis. *ACS Omega* 2021;6; 28455–28462. Available from: https://doi.org/10.1021/acsomega.1c03994

34. Guo L, Wang BW, Xu W, Yan R, Zhang Y, Bi W, et al. Engineered trimeric ACE2 binds viral spike protein and locks it in "three up" conformation to potently inhibit SARS-CoV-2 infection. *Cell Research* 2021;31; 98–100. Available from: https://doi.org/10.1038/s41422–020–00438-w

35. Hammond J, Leister-Tebbe H, Gardner A, Abreu P, Bao W, Wisemandle W, et al. Oral nirmatrelvir for high-risk, nonhospitalized adults with Covid-19. *New England Journal of Medicine* 2022;386; 1397–1408. Available from: https://doi.org/10.1056/NEJMoa2118542

36. Beheshtirouy S, Khani E, Khiali S, Entezari-Maleki T. Investigational antiviral drugs for the treatment of COVID-19 patients. *Archives of Virology* 2022;167; 751–805. Available from: https://doi.org/10.1007/s00705–022–05368-z

37. Novartis Pharmaceuticals. *Novartis Pharmaceuticals Phase 3 Multicenter, Randomized, Double-Blind, Placebo-Controlled Study to Assess the Efficacy and Safety of Canakinumab on Cytokine Release Syndrome in Patients with COVID-19-Induced Pneumonia (CAN-COVID); Clinical Trial Registration NCT04362813; Clinicaltrials.gov.* Tuebingen: University Hospital Tuebingen; 2021.

38. Federal Budgetary Research Institution State Research Center of Virology and Biotechnology "Vector". *Multicenter Double-Blind Placebo-Controlled Comparative Randomized Study of the Tolerability, Safety, Immunogenicity and Prophylactic Efficacy of the EpiVacCorona Peptide Antigen-Based Vaccine for the Prevention of COVID-19, with the Participation of 3000 Volunteers Aged 18 Years and Above (Phase III–IV); Clinical Trial Registration NCT04780035; Clinicaltrials.gov.* Tuebingen: University Hospital Tuebingen; 2021.

39. University Hospital Tuebingen. *P-PVAC-SARS-CoV-2: Phase I Single-Center Safety and Immunogenicity Trial of Multi-Peptide Vaccination to Prevent COVID-19 Infection in Adults; Clinical Trial Registration nct04546841clinicaltrials.gov.* Tuebingen: University Hospital Tuebingen; 2021.

40. Hamley IW. Peptides for vaccine development. *ACS Applied Bio Materials* 2022;5; 905–944. Available from: https://doi.org/10.1021/acsabm.1c01238

41. Deshayes C, Arafath MN, Apaire-Marchais V, Roger E. Drug delivery systems for the oral adminis-tration of antimicrobial peptides: Promising tools to treat infectious diseases. *Frontiers in Medical Technology* 2022;3; 778645. Available from: https://doi.org/10.3389/fmedt.2021.778645

42. Dijksteel G. S., Ulrich M. M. W., Middelkoop E., Boekema B. K. H. L. (2021). Review: Lessons learned from clinical trials using antimicrobial peptides (AMPs). *Frontiers in Microbiology* 2021;12; 616979. Available from: https://doi.org/10.3389/fmicb.2021.616979

43. Magana M., Pushpanathan M., Santos A. L., Leanse L., Fernandez M., Ioannidis A., et al. (2020). The value of antimicrobial peptides in the age of resistance. *Lancet Infectious Diseases* 20; e216–e230. Available from: https://doi.org/10.1016/S1473-3099(20)30327-3

44. Narusaka M, Hatanaka T, Narusaka Y. Inactivation of plant and animal viruses by proanthocyani-dins from *Alpinia zerumbet* extract. *Plant Biotechnology Journal* 2021;38; 453–455. Available from: https://doi.org/10.5511/plantbiotechnology.21.0925a.

45. Morimoto H, Hatanaka T, Narusaka M, Narusaka Y. Molecular investigation of proanthocyanidin from *Alpinia zerumbet* against the influenza a virus. *Fitoterapia* 2022;158; 105141. Available from: https://doi.org/10.1016/j.fitote.2022.105141

46. Mateos-Martín ML, Fuguet E, Jiménez-Ardón A, Herrero-Uribe L, Tamayo-Castillo G, Torres JL. Identification of polyphenols from antiviral *Chamaecrista nictitans* extract using high-resolution LC-ESI-MS/MS. *Analytical and Bioanalytical Chemistry* 2014;406; 5501–5506. Available from: https://doi.org/10.1007/s00216-014-7982-6.

47. Fink RC, Roschek B, Jr., Alberte RS. HIV type-1 entry inhibitors with a new mode of action. *Anti-viral Chemistry & Chemotherapy* 2009;19; 243–255. Available from: https://doi.org/10.1177/095632020901900604.

48. Connell BJ, Chang SY, Prakash E, Yousfi R, Mohan V, Posch W, Wilflingseder D, Moog C, Kodama EN, Clayette P, et al. A cinnamon-derived procyanidin compound displays anti-HIV-1 activity by blocking heparan sulfate- and co-receptor-binding sites on gp120 and reverses T cell exhaustion via impeding tim-3 and pd-1 upregulation. *PLoS One* 2016;11; e0165386. Available from: https://doi.org/10.1371/journal.pone.0165386.

49. Fauvelle C, Lambotin M, Heydmann L, Prakash E, Bhaskaran S, Vishwaraman M, Baumert TF, Moog C. A cinnamon-derived procyanidin type a compound inhibits hepatitis C virus cell entry. *Hepatology International* 2017;11; 440–445. Available from: https://doi.org/10.1007/s12072-017-9809-y.

50. Zhuang M., Jiang H., Suzuki Y., Li X., Xiao P., Tanaka T., Ling H., Yang B., Saitoh H., Zhang L., et al. Procyanidins and butanol extract of cinnamomi cortex inhibit SARS-CoV infection. *Antiviral Research* 2009;82; 73–81. Available from: https://doi.org/10.1016/j.antiviral.2009.02.001.

51. Tietjen I, Ntie-Kang F, Mwimanzi P, Onguéné PA, Scull MA, Idowu TO, Ogundaini AO, Meva'a LM, Abegaz BM, Rice CM, et al. Screening of the pan-African natural product library identifies ixoratannin a-2 and boldine as novel HIV-1 inhibitors. *PLoS One*. 2015;10; e0121099. Available from: https://doi.org/10.1371/journal.pone.0121099.

52. Xu X, Xie H, Wang Y, Wei X. A-type proanthocyanidins from lychee seeds and their antioxidant and antiviral activities. *Journal of Agricultural and Food Chemistry* 2010;58; 11667–11672. Available from: https://doi.org/10.1021/jf1033202.

53. Feng WY, Tanaka R, Inagaki Y, Saitoh Y, Chang MO, Amet T, Yamamoto N, Yamaoka S, Yoshinaka Y. Pycnogenol, a procyanidin-rich extract from French maritime pine, inhibits intracellular replication of HIV-1 as well as its binding to host cells. *Japanese Journal of Infectious Diseases* 2008;61; 279–285.

54. Ezzikouri S, Nishimura T, Kohara M, Benjelloun S, Kino Y, Inoue K, Matsumori A, Tsukiyama-Kohara K. Inhibitory effects of Pycnogenol® on hepatitis C virus replication. *Antiviral Research* 2015;113; 93–102. Available from: https://doi.org/10.1016/j.antiviral.2014.10.017

55. Suedee A, Tewtrakul S, Panichayupakaranant P. Anti-HIV-1 integrase compound from *Pometia pinnata* leaves. *Pharmaceutical Biology* 2013;51; 1256–1261. Available from: https://doi.org/10.3109/13880209.2013.786098.

56. Fink RC, Roschek B, Jr, Alberte RS. Hiv type-1 entry inhibitors with a new mode of action. *Antiviral Chemistry & Chemotherapy* 2009;19; 243–255. Available from: https://doi.org/10.1177/095632020901900604.

57. De Bruyne T, Pieters L, Witvrouw M, De Clercq E, Vanden Berghe D, Vlietinck AJ. Biological evalu-ation of proanthocyanidin dimers and related polyphenols. *Journal of Natural Products* 1999;62; 954–958. Available from: https://doi.org/10.1021/np980481o.

58. Luganini A, Terlizzi ME, Catucci G, Gilardi G, Maffei ME, Gribaudo G. The cranberry extract Oximacro(®) exerts in vitro virucidal activity against influenza virus by interfering with hemagglutinin. *Frontiers in Microbiology* 2018;9; 1826. Available from: https://doi.org/10.3389/fmicb.2018.01826.

59. Turmagambetova AS, Sokolova NS, Bogoyavlenskiy AP, Berezin VE, Lila MA, Cheng DM, Dushenkov V. New functionally-enhanced soy proteins as food ingredients with anti-viral activity. *Virusdiseae* 2015;26; 123–132. Available from: https://doi.org/10.1007/s13337-015-0268-6.

60. Sugamoto K, Tanaka YL, Saito A, Goto Y, Nakayama T, Okabayashi T, Kunitake H, Morishita K. Highly polymerized proanthocyanidins (PAC) components from blueberry leaf and stem significantly inhibit SARS-CoV-2 infection via inhibition of ACE2 and viral 3Clpro enzymes. *Biochemical and Biophysical Research Communications* 2022;615; 56–62. Available from: https://doi.org/10.1016/j.bbrc.2022.04.072.

61. Joshi SS, Howell AB, D'Souza DH. Reduction of enteric viruses by blueberry juice and blueberry proanthocyanidins. *Food and Environmental Virology* 2016;8; 235–243. Available from: https://doi.org/10.1007/s12560-016-9247-3.

62. Takeshita M, Ishida Y, Akamatsu E, Ohmori Y, Sudoh M, Uto H, Tsubouchi H, Kataoka H. Proanthocyanidin from blueberry leaves suppresses expression of subgenomic hepatitis C virus RNA. *Journal of Biological Chemistry* 2009;284; 21165–21176. Available from: https://doi.org/10.1074/jbc.M109.004945.

63. Lipson SM, Ozen FS, Karthikeyan L, Gordon RE. Effect of ph on anti-rotavirus activity by comestible juices and proanthocyanidins in a cell-free assay system. *Food and Environmental Virology* 2012;4; 168–178. Available from: https://doi.org/10.1007/s12560-012-9086-9.

64. Nair M.P., Kandaswami C., Mahajan S., Nair H.N., Chawda R., Shanahan T., Schwartz S.A. Grape seed extract proanthocyanidins downregulate hiv-1 entry coreceptors, ccr2b, ccr3 and ccr5 gene expression by normal peripheral blood mononuclear cells. *Biological Research* 2002;35; 421–431. Available from: https://doi.org/10.4067/S0716-97602002000300016.

65. Marin-Hernandez D, Hupert N, Nixon DF. The immunologists' guide to pandemic preparedness. *Trends in Immunology* 2021;42; 91–93.

66. Chambers ES, et al. Vitamin D3 replacement enhances antigen-specific immunity in older adults. *Immunotherapy Advances* 2021; 1.

67. Opatowski L, Baguelin M, Eggo RM. Influenza interaction with cocirculating pathogens and its impact on surveillance, pathogenesis, and epidemic profile: A key role for mathematical modelling. *PLOS Pathogens* 2018;14; e1006770.

68. Nickbakhsh S, et al. Virus-virus interactions impact the population dynamics of influenza and the common cold. *Proceedings of the National Academy of Sciences of the United States of America* 2019 Dec 26;116(52); 27142–27150.

69. Dianzani F. Viral interference and interferon. *Ric Clin Lab*. 1975;5; 196–213.

70. Marin-Hernandez D, Hupert N, Nixon DF. The immunologists' guide to pandemic preparedness. *Trends Immunology* 2021;42; 91–93.

71. Hupert N, Marin-Hernandez D, Nixon DF. Can existing unrelated vaccines boost a COVID-19 vaccine prime? *EClinicalMedicine* 2021;32; 100758.

72. Moorlag S, Arts RJW, van Crevel R, Netea MG. Non-specific effects of BCG vaccine on viral infections. *Clin Microbiol Infect*. 2019;25; 1473–1478.

73. Wrangle JM, et al. ALT-803, an IL-15 superagonist, in combination with nivolumab in patients with metastatic non-small cell lung cancer: a non-randomised, open-label, phase 1b trial. *Lancet Oncol*. 2018;19; 694–704.

74. Rosenberg SA. IL-2: The first effective immunotherapy for human cancer. *J Immunol*. 2014;192; 5451–5458.

75. Long BR, et al. Elevated frequency of gamma interferon-producing NK cells in healthy adults vaccinated against influenza virus. *Clin Vaccine Immunol*. 2008;15; 120–130.

76. Ram DR, Manickam C, Lucar O, Shah SV, Reeves RK. Adaptive NK cell responses in HIV/SIV infections: A roadmap to cell-based therapeutics? *J Leukoc Biol*. 2019;105; 1253–1259.

77. Market M, et al. Flattening the COVID-19 curve with natural killer cell based immunotherapies. *Front Immunol*. 2020;11; 1512.

78. Ram DR, Manickam C, Lucar O, Shah SV, Reeves RK. Adaptive NK cell responses in HIV/SIV infections: A roadmap to cell-based therapeutics? *J Leukoc Biol*. 2019;105; 1253–1259.

79. Poccia F, et al. Vgamma9Vdelta2 T cell-mediated non-cytolytic antiviral mechanisms and their potential for cell-based therapy. *Immunol Lett*. 2005;100; 14–20.

80. Pistoia V, et al. Human γδ T-cells: From surface receptors to the therapy of high-risk leukemias. *Front Immunol*. 2018;9; 984.

81. Wakao H, Sugimoto C, Kimura S, Wakao R. Mucosal-associated invariant T cells in regenerative medicine. *Front Immunol*. 2017;8; 1711.

82. La Manna MP, et al. Harnessing unconventional T cells for immunotherapy of tuberculosis. *Front Immunol*. 2020;11; 2107.

83. Wang Z, et al. MR1-restricted T cells: The new dawn of cancer immunotherapy. *Biosci Rep*. 2020; 40.

84. Marin-Hernandez D, Hupert N, Nixon DF. The immunologists' guide to pandemic preparedness. *Trends Immunol*. 2021;42; 91–93.

85. Calabrese LH, Lenfant T, Calabrese C. Interferon therapy for COVID-19 and emerging infections: Prospects and concerns. *Cleve Clin J Med*. 2020.

86. Brzoska J, von Eick H, Hundgen M. Interferons in the therapy of severe coronavirus infections: a critical analysis and recollection of a forgotten therapeutic regimen with interferon beta. *Drug Res (Stuttg)*. 2020;70; 291–297.

87. Zylberman V, et al. Development of a hyperimmune equine serum therapy for COVID-19 in Argentina. *Medicina (B Aires)*. 2020;80(Suppl 3); 1–6.

88. Long BR, et al. Elevated frequency of gamma interferon-producing NK cells in healthy adults vaccinated against influenza virus. *Clin Vaccine Immunol*. 2008;15; 120–130.

89. Kumar RN et al. Real-world experience of bamlanivimab for COVID-19: A case-control study. *Clin Infect Dis*. 2022 Jan 7;74(1); 24–31.

90. Boskovic M, Migo W, Likic R. SARS-CoV-2 mutations: A strain on efficacy of neutralizing monoclonal antibodies? *Br J Clin Pharmacol*. 2021 Nov;87(11); 4476–4478.

91. Lucas C, et al. Delayed production of neutralizing antibodies correlates with fatal COVID-19. *Nat Med*. 2021, Jul;27(7); 1178–1186.

92. Honjo K, et al. Convalescent plasma-mediated resolution of COVID-19 in a patient with humoral immunodeficiency. *Cell Rep Med*. 2021;2; 100164.

93. Honjo K, et al. Convalescent plasma-mediated resolution of COVID-19 in a patient with humoral immunodeficiency. *Cell Rep Med*. 2021;2; 100164.

94. Keller MD, et al. SARS-CoV-2-specific T cells are rapidly expanded for therapeutic use and target conserved regions of the membrane protein. *Blood*. 2020;136; 2905–2917.

95. McCann CD, et al. A participant-derived xenograft model of HIV enables long-term evaluation of autologous immunotherapies. *J Exp Med*. 2021; 218.

96. Bruxvoort KJ, et al. Patient report of herpes zoster pain: Incremental benefits of zoster vaccine live. *Vaccine*. 2019;37; 3478–3484.

97. Dangi AK., Sinha R, Dwivedi S, Gupta SK, Shukla P. Cell line techniques and gene editing tools for antibody production: A review. *Front. Pharmacol*. 2018;9; 630. Available from: https://doi.org/10.3389/fphar.2018.00630

98. Vashistha R, Chhabra D, Shukla P. Integrated artificial intelligence approaches for disease diagnostics. *Indian J. Microbiol*. 2018;58 (2); 252–255. Available from: https://doi.org/10.1007/s12088-018-0708-2

99. Metsky HC, Freije CA, Kosoko-Thoroddsen TSF, Sabeti PC, Myhrvold C. CRISPR-based COVID-19 surveillance using a genomically-comprehensive machine learning approach. *bioRxiv* 2020. Available from: https://doi.org/10.1101/2020.02.26.967026

100. Rahmatizadeh S, Valizadeh-Haghi S, Dabbagh A. The role of artificial intelligence in management of critical COVID-19 patients. *J. Cell. Mol. Anesthesia*. 2020;5(1); 16–22.

101. Wu C, Chen X, Cai Y, Zhou X, Xu S, Huang H, et al. Risk factors associated with acute respiratory distress syndrome and death in patients with coronavirus disease 2019 pneumonia in Wuhan, China. *JAMA Internal Med*. 2020a. Available from: https://doi.org/10.1001/jamainternmed.2020.0994

102. Alimadadi A, Aryal S, Manandhar I, Munroe PB, Joe B, Cheng X. Artificial intelligence and machine learning to fight COVID-19. *Physiol Genomics*. 2020;52; 200–202. Available from: https://doi.org/10.1152/physiolgenomics.00029.2020

103. Hu X, Go YM, Jones DP. Omics integration for mitochondria systems biology. *Antioxid. Redox Signaling*. 2020;32(12); 853–872. Available from: https://doi.org/10.1089/ars.2019.8006

104. Lan J, Ge J, Yu J, Shan S, Zhou H, Fan S, et al. Structure of the SARS-CoV-2 spike receptor-binding domain bound to the ACE2 receptor. *Nature* 2020a; 1–9. Available from: https://doi.org/10.1101/2020.02.19.956235

105. Lan J, Ge J, Yu J, Shan S, Zhou H, Fan S, et al. Crystal structure of the 2019-nCoV spike receptor-binding domain bound with the ACE2 receptor. *bioRxiv* 2020b. Available from: https://doi.org/10.1101/2020.02.19.956235

106. Chen J. Pathogenicity and transmissibility of 2019-nCoV—a quick overview and comparison with other emerging viruses. *Microbes Infect.* 2020. Available from: https://doi.org/10.1016/j.micinf.2020.01.004

107. Nasi A, McArdle S, Gaudernack G, Westman G, Melief C, Arens R, et al. Proteasome and reactive oxygen species dysfunction as risk factors for SARS-CoV infection; consider N-acetylcystein as therapeutic intervention. *Toxicol. Rep.* 2020;7; 768–771.

108. van de Laar L, Saelens W, De Prijck S, Martens L, Scott CL, Van Isterdael G, et al. Yolk sac macrophages, fetal liver, and adult monocytes can colonize an empty niche and develop into functional tissue-resident macrophages. *Immunity* 2016;44(4); 755–768. Available from: https://doi.org/10.1016/j.immuni.2016.02.017

109. Cousins KR. *Computer Review of Chemdraw Ultra 12.0.* Washington, DC: ACS Publications; 2011.

110. Waterhouse A, Bertoni M, Bienert S, Studer G, Tauriello G, Gumienny R, et al. SWISS-MODEL: Homology modelling of protein structures and complexes. *Nucleic Acids Res.* 2018;46(W1); Available from: W296–W303. https://doi.org/10.1093/nar/gky427

111. Colovos C, Yeates TO. Verification of protein structures: Patterns of nonbonded atomic interactions. *Protein Sci.* 1993;2(9); 1511–1519. Available from: https://doi.org/10.1002/pro.5560020916

112. Wiederstein M., Sippl M. J. (2007). ProSA-web: Interactive web service for the recognition of errors in three-dimensional structures of proteins. *Nucleic Acids Res.* 2007; 35(Suppl 2); W407–W410. Available from: https://doi.org/10.1093/nar/gkm290

113. Wallner B. *Protein Structure Prediction: Model Building and Quality Assessment* Doctoral dissertation, Institution en för Biokemioch Biofysik; 2005.

114. Lovell SC, Davis IW, Arendall WB, de Bakker PII, Word JM, Prisant MG, et al. Structure validation by Calpha geometry: Phi, psi and Cbeta deviation. *Proteins* 2003;50; e437–e450. Available from: https://doi.org/10.1002/prot.10286

115. Dromey RG. Cornering the chimera [software quality]. *IEEE Software* 1996;13(1); 33–43. Available from: https://doi.org/10.1109/52.476284

116. DeLano WL. Pymol: An open-source molecular graphics tool. CCP4 Newslett. *Protein Crystallogr.* 2002;40(1); 82–92.

117. Norgan AP, Coffman PK, Kocher JPA, Katzmann DJ, Sosa CP. Multilevel parallelization of AutoDock 4.2. *J. Cheminform.* 2011;3(1); 12. Available from: https://doi.org/10.1186/1758-2946-3-12

118. Schrodinger L. Schrodinger Software Suite. New York: Schrödinger, LLC; 2011, 670.

119. Avadhanula V, Wang Y, Portner A, Adderson E. Nontypeable *Haemophilus influenzae* and *Streptococcus pneumoniae* bind respiratory syncytial virus glycoprotein. *J Med Microbiol* 2007;56(Pt. 9); 1133–1137.

120. Avadhanula V, Rodriguez CA, DeVincenzo JP, Wang Y, Webby RJ, Ulett GC, et al. Respiratory viruses augment the adhesion of bacterial pathogens to respiratory epithelium in a viral species- and cell type-dependent manner. *J Virol.* 2006;80(4); 1629–1636.

121. Taylor DL, Kahawita TM, Cairncross S, Ensink JH. The impact of water, sanitation and hygiene interventions to control cholera: A systematic review. *PLoS One* 2015;10(8); e0135676. Available from: https://doi.org/10.1371/journal.pone.0135676. PMID: 26284367; PMCID: PMC4540465.

122. Wilson AL, Courtenay O, Kelly-Hope LA, Scott TW, Takken W, Torr SJ, Lindsay SW. The importance of vector control for the control and elimination of vector-borne diseases. *PLoS Negl Trop Dis.* 2020;14(1); e0007831. Available from: https://doi.org/10.1371/journal.pntd.0007831. PMID: 31945061; PMCID: PMC6964823.

123. Hammond AM, Galizi R. Gene drives to fight malaria: Current state and future directions. *Pathog Glob Health.* 2017;111; 412–423. Available from: https://doi.org/10.1080/20477724.2018.1438880

124. Flores HA, O'Neill SL. Controlling vector-borne diseases by releasing modified mosquitoes. *Nat Rev Microbiol.* 2018;16(8); 508–518. Available from: https://doi.org/10.1038/s41579-018-0025-0

125. Knols BG, Farenhorst M, Andriessen R, Snetselaar J, Suer RA, Osinga AJ, et al. Eave tubes for malaria control in Africa: An introduction. *Malar J.* 2016;15; 404. Available from: https://doi.org/10.1186/s12936-016-1452-x

Index

Note: Page numbers in *italics* indicate a figure and page numbers in **bold** indicate a table on the corresponding page.

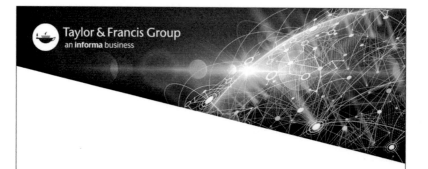